Urban Health and Wellbeing

Systems Approaches

Series Editor

Yongguan Zhu, Chinese Academy of Sciences, Institute of Urban Environment, Xiamen, Fujian, China

The Urban Health and Wellbeing: Systems Approaches series is based on a 10-year global interdisciplinary research program developed by International Council for Science (ICSU), and sponsored by the InterAcademy Partnership (IAP) and the United Nations University (UNU). It addresses up-to-date urban health issues from around the world and provides an appealing integrated urban development approach from a systems perspective. This series aims to propose a new conceptual framework for considering the multi-factorial and cross sectorial nature of both determinants and drivers of health and Wellbeing in urban populations and takes a systems approach for improving the understanding of the interconnected nature of health in cities. The systems approach includes an engagement with urban communities in the process of creating and transferring knowledge. Further, it aims at generating knowledge and providing the evidence that is relevant to people and policy-makers for improving integrated decision making and governance for the health and Wellbeing of urban dwellers. The methods applied, come from various epistemological domains in order to improve understanding of how the composition and functioning of urban environments impacts physical, mental and social heath and how inequalities can be reduced to improve the overall quality of urban life.

The systems approach is applied to science and society and defined by a deep investigation into disciplinary knowledge domains relevant for urban health and Wellbeing, as well as an inter- and transdisciplinary dialogue and shared understanding of the issues between scientific communities, policy makers and societal stakeholders more broadly. It involves one or more of the following elements: 1) the development of new conceptual models that incorporate dynamic relations among variables which define urban health and wellbeing; 2) the use of systems tools, stimulation models and collaborative modelling methods; 3) the integration of various sources and types of data including spatial, visual, quantitative and qualitative data.

Like the first book, the coming books will all address the topic of urban health and wellbeing, specifically by taking a systems approach. The topics range across all urban sectors and can, for example, cover the following areas:

(1) transportation, urban planning and housing, urban water, energy and food, communication, resources and energy, urban food systems, public service provision, etc.
(2) the related health disorders in physical, social and mental health
(3) the methods and models used and the type of science applied to understand the complexity of urban health and wellbeing.

More information about this series at http://www.springer.com/series/15601

Franz W. Gatzweiler

Editor

Urban Health and Wellbeing Programme

Policy Briefs: Volume 1

Editor
Franz W. Gatzweiler
Institute of Urban Environment (IUE)
Chinese Academy of Sciences (CAS)
Xiamen, China

ISSN 2510-3490   ISSN 2510-3504   (electronic)
Urban Health and Wellbeing
ISBN 978-981-15-1382-4   ISBN 978-981-15-1380-0   (eBook)
https://doi.org/10.1007/978-981-15-1380-0

Jointly published with Zhejiang University Press
The print edition is not for sale in China Mainland. Customers from China Mainland please order the print book from: Zhejiang University Press.

This Springer imprint is published by the registered company Springer Nature Singapore Pte Ltd.
The registered company address is: 152 Beach Road, #21-01/04 Gateway East, Singapore 189721, Singapore

Preface

Cities are the new melting pots of global development. Over half of the world's population lives in cities, and this number is increasing by about 2% annually. More than two billion urban dwellers are expected to be added over the next three decades, a significant proportion of whom will be living in informal or slum settlements. Urban areas are extremely complex environments in which environmental, social, cultural and economic factors influence people's health and wellbeing.

The Urban Health and Wellbeing Programme (UHWB) is a global science programme and interdisciplinary body of the International Science Council (ISC, previously ICSU), supported by the United Nations University (UNU), the InterAcademy Partnership (IAP) and the International Society on Urban Health. The overarching vision for UHWB is people to develop aspired levels of wellbeing by living in healthy cities.

This book is a compilation of the policy briefs published by the UHWB programme over the past three years, which aims at highlighting and drawing attention to policy-relevant findings and insights from research and researchers and communicating them with decision-makers at all levels of society in order to encourage the co-creation of knowledge for healthy urban environments and people. It shows how system approaches promoted by the programme have been applied widely by the world and contributed to global urban health.

Xiamen, China   Franz W. Gatzweiler

Contents

A Systems Approach to Urban Health and Well-being Has Come of Age in China

Xinhu Li

Key Messages

1. Ecological civilization is no longer only a conceptual innovation but also a key national governance strategy in China.
2. The State Council released the worldwide first national strategy for the circular economy, and the circular economy became a national development strategy in China.
3. The "Healthy China 2030" Plan issued by the CPC Central Committee and the State Council confirmed the central position of the "Healthy China Plan" to the Chinese Government's agenda for health and development.
4. China has the biggest population in the world and has been experiencing the largest migration in history. Its rapid urbanization has profound and lasting impacts on local, national, and international public health.
5. While at the individual level, people choose behaviors or living environments which relate to their own health, the option for improving well-being should be provided by the government.

1 Ecological Civilization

The Chinese Government has been paying attention to ecological and environmental issues for many years. In 1983, environmental protection was identified as a basic national policy during the Second Environmental Protection Working Conference. Sustainable development was established as a national strategy after the 15th CPC National Congress in 1997. Ecological civilization was proposed for the first time

X. Li (✉)
College of Architecture and Urban Planning, Tongji University, Shanghai, China
e-mail: xhli@tongji.edu.cn

© Zhejiang University Press and Springer Nature Singapore Pte Ltd. 2020   1
F. W. Gatzweiler (ed.), *Urban Health and Wellbeing Programme*,
Urban Health and Wellbeing, https://doi.org/10.1007/978-981-15-1380-0_1

during the 17th CPC National Congress in 2007 and elevated as a political outline and national strategy of governance during 18th CPC in 2012.

On September 21, 2015, the CPC Central Committee and State Council issued the "Overall Plan for the Reform of Ecological Civilization System" which defined the overall design and roadmap for an ecological civilization in the future. In 2016, the concept of green development was proposed and ecological civilization was incorporated into the 13th Five-Year Plan. The introduction of all these major policies means that ecological civilization is no longer only a conceptual innovation but also a key national governance strategy for China. Furthermore, under the influence of this strategy, a series of practical actions have been implemented. To achieve the green development target, the government has taken a series of actions for intensive resource use and energy conservation, including the water regulations, farmland protection regulations, and environmental regulations in addition to promoting energy savings and industrial structural adjustment.

2 Circular Economy

More than ten years ago, China's central government recognized the economic and environmental risk of the nation's heavy resource exploitation and adopted the circular economy as the principal means of dealing with these problems. The circular economy is a system in which the use and waste of resources and energy are minimized by closing material and energy cycles. In 2005, China's State Council issued a policy paper called "Several Opinions of the State Council on Accelerating the Development of Circular Economy." According to this document, the Development and Reform Commission, together with relevant departments such as the Environmental Protection Administration and other relevant departments, should carry out supervision and inspection of the development of the circular economy and report to the State Council.

A series of taxation, fiscal, pricing, and industrial policies were introduced to promote the development of the circular economy. A fund was set up to support the conversion of industrial parks into eco-industrial agglomerations. In the 11th Five-Year Plan (2006–2010), a whole chapter was devoted to the circular economy. In 2008, The Fourth Session of the Standing Committee of the 11th National People's Congress of the People's Republic of China adopted the Circular Economy Promotion Law, which demanded that local governments consider circular economy in their investment and development strategies. In the 12th Five-Year Plan (2011–2015), the circular economy was upgraded to a national development strategy. The State Council released the first national strategy for circular economy in the world in 2013.

In 2016, the General Office of the State Council issued the "extension of the producer responsibility system" to implement the program. Electrical and electronic products, automotive products, lead-acid batteries, and packaging products were chosen as pilot categories. The implementation of the producer responsibility extension system will help to build a circular waste disposal system and to make use of the long-term promotion of the ecological civilization.

3 Healthy China 2030

As early as September 08, 2007, at the annual meeting of the China Association for Science and Technology, Minister of Health Chen Zhu announced the three-step strategy of "health protection and well-being well-off, well-off rely on health," and revealed the relevant action plan. This strategic plan aims to achieve three objectives: (1) the initial establishment of the basic health-care system framework for urban and rural residents by 2010, (2) to develop China's health-care system to the forefront of developing countries by 2015, and (3) by 2020, to maintain the status of Healthy China at the forefront of the developing countries, and the eastern part of the urban and rural areas and some of the urban and rural areas close to or reach the level of middle developed countries.

On August 17, 2012, the Ministry of Health issued the "Healthy China 2020 Strategic Research Report." It proposed eight health-related policy recommendations. In 2015, in his report to the government, Premier Li Keqiang first proposed "to build a healthy China." The Fifth Plenary Session of the Eighth Central Committee of the Party further put forward the task requirements of "promoting the building of Healthy China." In August 2016, President Xi said at the National Health Conference that health was a prerequisite for people's overall development and a precondition for economic and social development. Following the National Health Conference, in October 2016, the CPC Central Committee and the State Council issued the "Healthy China 2030" Plan and issued a notice calling on all localities and departments to implement the plan, which confirmed the central position of Healthy China 2030 Plan to the Chinese Government's agenda for health and development. This document is the first medium to long-term strategic plan in the health sector developed at the national level since the founding of China in 1949.

4 Case Studies

The research group of Dr. Li Xinhu at the Institute of Urban Environment systematically analyzes the ways and factors influencing the health of the population affected by urbanization in China and puts forward relevant strategies and policy suggestions from the national, local, and individual levels. Figure 1 illustrates how systemic changes in the environment as a result of urbanization pose numerous threats to human health. Rapid and often unplanned, urban growth is a source of environmental hazards, which have direct and indirect effects on human health.

Urban expansion is one of the major driving factors of land use change in China, with extensive effects on local ecological systems through reducing biodiversity, air

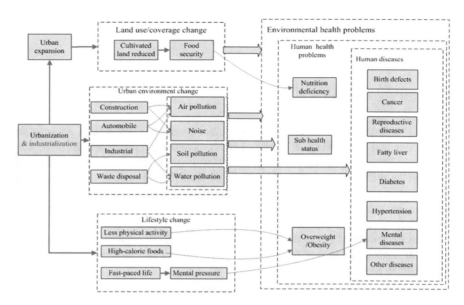

Fig. 1 Relationship between urbanization, urban environmental change, and public health

deterioration and contributing to water shortages. Accelerated urbanization along with explosive economic growth has further worsened the shortage of agricultural land over the last two decades with possible consequences for food security and nutritional deficiencies threatening the overall health status of the population.

Reduced cultivated land places pressures to intensify agricultural production which depends on both the progress of agricultural technology and higher use of fertilizer and pesticides. Such inputs have repercussions for the availability of safe food, and also for the price of food, as fertilizer costs increase in line with oil prices. A multi-level understanding of the relationship between the features of China's urbanization, urban environmental change, and risks to public health provides the basis for identifying interventions at national, local, and individual levels (Li et al. 2016).

5 Policy Recommendations

Based on the evidence provided by our research group, attention needs to focus on the following key public health domains:

1. a pollution-free environment
2. a safe and diverse food supply
3. a health system that addresses the needs of hard-to-reach groups
4. planning for healthy cities
5. health behavior education.

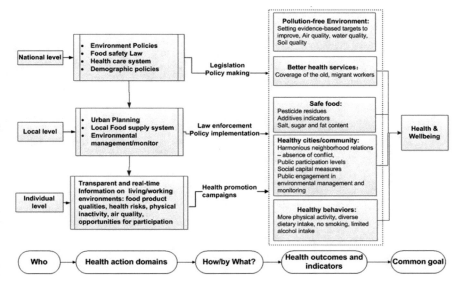

Fig. 2 Choices of health-promoting behavior at individual level and options for improving wellbeing at governmental level

While at the individual level, people choose behaviors or living environments which relate to their own health, the option for improving well-being should be provided by the government (Fig. 2). Coordinated action by the central government, the local government, and the public in each of these areas could advance the goal of health and well-being for all of the citizens in China. The figure below describes the major categories of health-promoting activity alongside the administrative or governance level which has major responsibility for the activity, the types of regulatory or change mechanism which would be involved, and examples of the types of measurable indicators of achievement toward the shared goal of improved health and well-being (Li et al. 2016).

Reference

Li X, Song J, Lin T et al (2016) Urbanization and health in China, thinking at the national, local and individual levels. Environ. Health 15(S1)

A Systems Approach to Urban Health and Well-being Has Come of Age in the Asia-Pacific Region

José Siri and David Tan

Key Messages

Experiences with a systems approach to urban health and well-being in the Asia-Pacific region underline the need to:

1. Understand and underscore the centrality of health in development by formalizing links between health and other sectors.
2. Move beyond simple indicators to a recognition of the consequences of complexity, and particularly of the dynamics of causal feedback loops.
3. Identify important cross-sectoral synergies and trade-offs.
4. Promote inter- and transdisciplinarity in science through new funding and evaluation criteria.
5. Create mechanisms and structures to improve the science/policy interface.

1 Policy Context

The Asia-Pacific region (Fig. 1), home to more than half of Earth's population (UN-DESA 2017), is experiencing urban growth at unprecedented scales; urban land, for example, increased by > 22% (equivalent to the area of Taiwan) and urban populations by > 31% in East-Southeast Asia in just a decade from 2000 to 2010 (Schneider et al. 2015).

This extraordinary expansion entails a parallel increase in complexity, a natural consequence of the proliferation of governance entities needed to deal with growing

J. Siri (✉)
Senior Science Lead for Cities and Health, Wellcome Trust, London, UK
e-mail: J.Siri@wellcome.ac.uk

D. Tan
The United Nations University, International Institute for Global Health, Kuala Lumpur, Malaysia

© Zhejiang University Press and Springer Nature Singapore Pte Ltd. 2020   7
F. W. Gatzweiler (ed.), *Urban Health and Wellbeing Programme*,
Urban Health and Wellbeing, https://doi.org/10.1007/978-981-15-1380-0_2

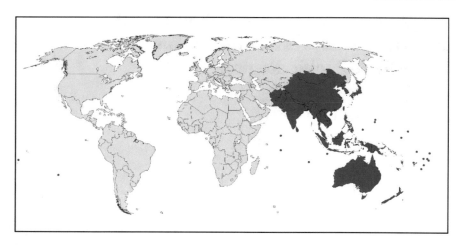

Fig. 1 The asia-pacific region. Created on: www.mapchart.net

cities, the multiplying interconnections between elements of urban systems (e.g., people, firms, infrastructure, and institutions), the physical expansion of urban areas, and advances in technology that magnify the impacts of individual actors (e.g., telecommunications, transport), among other factors. Urban complexity involves interlinked causal feedback loops that are rarely perceived in whole by those involved. As such, policy decisions have numerous impacts beyond their intended effects; the former often outweigh the latter, leading to policy surprise or failure (Newell and Proust 2018). Moreover, because impacts can have significant delays, and because all actions are taken in the context of many simultaneous decisions, it is difficult to evaluate the efficacy of any particular policy or intervention.

Complexity is thus a significant factor in adverse urban health outcomes. For example, the growing prevalence of non-communicable diseases (NCDs) is at least in part a consequence of the complexity of their environmental, behavioral, and physiological causes, which pose challenges for prevention and control (Lee et al. 2017). NCDs are now the most significant cause of deaths and disease burden in the Asia-Pacific region (Low et al. 2015). Changes in agricultural, behavioral, and ecological dynamics and human–animal interactions in urban areas also imply complex new infectious disease risks (Hassell et al. 2017)—indeed, urban expansion in Asia has been tied to higher risk of highly pathogenic avian influenza (Saksena et al. 2017), and most new influenza subtypes and strains of seasonal influenza have originated in the region (Wen et al. 2016).

The unique challenges of Asian urban expansion are driving novel approaches to urban planning and management (see, e.g., Baculinao 2017); it is critical that such approaches account for complexity as it relates to health outcomes. Systems approaches are a set of actions intended to improve decision-making in such contexts. Fundamentally, they consist of two related elements: (a) analytic methodologies to uncover feedback relationships and other nonlinear elements of causal systems; and

(b) broad processes of engagement (interdisciplinary, transdisciplinary, and multi-scale) to improve problem characterization and ensure feasibility and buy-in (Siri 2016).

The need for systems approaches to complex problems, including in urban health, is implicit in the ever more integrated nature of platforms for sustainable development, most notably the Sustainable Development Goals (SDGs). It has been explicitly recognized in the academic literature (see, e.g., Bai et al. 2016; Newell and Siri 2016; Yang et al. 2018; Ramaswami et al. 2016) and is beginning to emerge in national and regional policy documents. The following section highlights an example of the application of systems approaches in Malaysia—the SCHEMA project—and offers a set of basic policy recommendations.

2 The SCHEMA Project

Since 2016, the United Nations University International Institute for Global Health (UNU-IIGH) and the Cardiff University Sustainable Places Research Institute (CU-PLACE) have co-led the SCHEMA project to promote more effective decision-making for urban health and sustainability in Malaysian cities: "Systems Thinking and Place-Based Methods for Healthier Malaysian Cities," funded by the British Council's Newton-Ungku Omar Fund.

SCHEMA draws on the Collaborative Conceptual Modeling (CCM) framework developed by Newell and Proust (2018) at Australian National University (Fig. 2), which uses simple system dynamics models to help participants visualize systemic structures and nonlinear relationships, and to improve communication and generate shared understanding (Newell and Siri 2016).

Fig. 2 Collaborative conceptual modeling framework. *Source* Newell and Proust (2018)

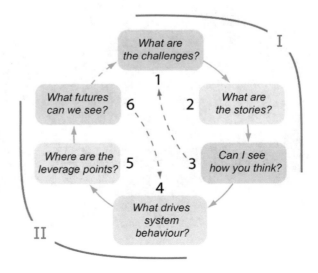

At the core of the SCHEMA project is an effort to develop a community of practice for systems approaches across the Malaysian government, academic, civil society, and private sectors, and to enable and encourage this transdisciplinary group to apply such approaches jointly in addressing health and sustainability challenges. To this end, it has applied various methodologies.

First, the project implemented a series of transdisciplinary workshops focused on methodological capacity building, cross-domain information exchange in thematic areas relevant to Malaysian cities—including green infrastructure, food systems, and campus sustainability—and exploration of structural and mechanistic factors that affect outcomes. These workshops were oriented toward the development of concrete proposals for new research collaborations, urban interventions, and organizational structures and mechanisms.

Second, the project led the co-production, with external authors from the SCHEMA community, of a series of case studies (SCHEMA 2018a) on complex urban problems related to health and/or sustainability. These case studies, developed through an iterative process of exchange between UNU researchers and outside partners, are simultaneously a capacity-building tool (i.e., for the use of causal loop diagrams and participatory model building) and an illustration of the value of systems methods for revealing fundamental causes, the potential impact of contrasting perspectives, and leverage points for intervention.

Box 1: Selected Examples of Systems Approaches for Urban Health in the Asia-Pacific

- Singapore-based Zeroth Labs blends methods from anthropology, data analysis, human-centered design, system dynamics, scenario planning and behavioral science to co-create new forms of public services or citizen solutions. They work "across a range of social issues, including sanitation, waste management, youth, education, and healthcare." (Chandran 2017).
- Through the Coastal Cities at Risk (CCAR) project's work in metropolitan Manila in the Philippines, "systems thinking about climate change adaptation have been mainstreamed in national and local government processes through specific policy instruments in collaboration with the all-levels of government, the military, regional scientific groups … and private sector partners." (McBean et al. 2017).
- Victoria, Australia's Food-Sensitive Planning and Urban Design report took an implicitly systems-based approach, "… draw[ing] on insights from academia, governance and practice in the disciplines of planning, urban design, sustainability and health," considering all phases of the food system and the roles and competing objectives of a diversity of stakeholders (Donovan et al. 2011).

- Since 2006, the Disha India Education Foundation has worked with over 50 schools across India to "...integrat[e] systems thinking principles and habits in the school curriculum and pedagogy," and promote experiential learning.
- The Tsinghua–Lancet Commission Report on Healthy Cities in China highlights the need for participatory systems-based efforts in developing China's urban future, explicitly calling for "transdisciplinary, interconnected, integrated and inclusive approaches—systems approaches—to deal with complex health challenges." (Yang et al. 2018).

Third, SCHEMA developed and disseminated the outputs to promote systemic thinking about urban complexity among lay people. For example, the project co-sponsored a photo competition with Think City, a Malaysian community-based urban regeneration organization, to inspire recognition of linkages among the Sustainable Development Goals. This was published as "THRIVE Connect: Linkages in Everyday Life," (SCHEMA 2018b) and launched at the 9th World Urban Forum (WUF9) in Kuala Lumpur. In the same vein, project researchers brought systems thinking messages to the public through a series of seminars, public talks, and exhibitions, leveraging international events such as WUF9 and platforms provided by Think City and other policy, research, and academic partners. These presentations drew on common experiences in urban life, demonstrating how these shape and are shaped by larger social–ecological systems.

Finally, the project has produced focused research on topics relevant to urban health and sustainability in the Malaysian context [e.g., the adoption of climate-sensitive buildings to mitigate the health impacts of urban heat (Tan et al. 2017)], drawing on the same systems methods promoted throughout.

The SCHEMA experience has demonstrated the potential for systems approaches to aid in developing and communicating a holistic understanding of complex urban health challenges. It has also revealed the need for institutional structures and mechanisms to enable the cross-sector interfacing required to fully utilize these understandings.

SCHEMA is but one of many initiatives in the Asia-Pacific now adopting and adapting systems approaches for urban health (See Box 1). Moreover, incipient large-scale efforts like China's Belt and Road Initiative and India's Smart Cities Mission offer possibilities for much wider adoption of systems approaches across the Asia-Pacific region, as does the broad program of action sponsored by the Asian Development Bank.

3   Policy Recommendations

Drawing on experiences from the SCHEMA project and other systems-based work in the region, we suggest the following concrete recommendations for actors working to improve urban health in the Asia-Pacific:

- Understand and underscore the centrality of health in development, adopting formal requirements for the cross-sectoral engagement. The World Health Organization has emphasized that effective health promotion requires action across all sectors (WHO 2014) and that good health is a precondition for the achievement of sustainable urban development (WHO 2016). Effective health messaging can powerfully motivate action, and in many contexts, health sector representatives (e.g., doctors) are trusted voices. Conversely, "non-health" expertise, including traditional and local community knowledge, has much to offer to the design and implementation of public health interventions.
- Move beyond simple indicators. While straightforward quantitative indicators are critical for benchmarking progress, decision-makers should at the same time strive to identify the important feedback loops, especially cross-sectoral feedbacks, that underlie health challenges. Thinking through feedback relationships will help actors to articulate theories of change and the expected pathways along which change will occur. In this manner, deviations from expectations can be more readily monitored, recognized, and addressed.
- Identify important cross-sectoral synergies and trade-offs. The International Council for Science recently published a framework for evaluating interactions among SDG targets (ICSU 2017), which can be viewed as a preliminary step toward a systems-based approach to sustainable development. At a time when many resources are growing scarce and increasing recognition of anthropogenic pressure on planetary systems dictates that others be conserved, it is especially critical to adopt actions that yield co-benefits and do not work counter to their own purposes.
- Promote inter and transdisciplinary in science, including new funding and new criteria for evaluation of scientific proposals and outputs. Large-scale initiatives in the region, such as China's Belt and Road Initiative and India's Smart Cities Mission, along with major development actors like the Asian Development Bank, have the opportunity to change the conversation by adopting systems thinking as a working paradigm.
- Create mechanisms/structures to improve the science/policy interface. There is a need for ongoing, long-term, close engagement between academics and policy-makers, a need that cannot be met solely through the production of policy-oriented documents as per standard practice, or by ad hoc one-off meetings dedicated to "knowledge transfer." One avenue to facilitate such engagement would be the promotion of systems thinking in secondary education; priming future generations to receive and engage in systems-based messages has the potential to significantly improve decision-making for health. A second avenue is participatory budgeting, which establishes a formal ongoing mechanism for policy–community interactions.

References

Baculinao E (2017) China plans a city the size of New England that'll be home to 130 Million. NBC News

Bai X, Surveyer A, Elmqvist T, Gatzweiler FW, Güneralp B, Parnell S, Prieur-Richard A-H et al (2016) Defining and advancing a systems approach for sustainable cities. Curr Opin Environ Sustain Open Issue Part I 23(December):69–78

Chandran N (2017) This start-up consultancy is using anthropology, data and design thinking to disrupt

Donovan J, Larsen K, McWhinnie J (2011) Food-sensitive planning and urban design: a conceptual framework for achieving a sustainable and healthy food system. National Heart Foundation of Australia (Victorian Division), Melbourne

Hassell JM, Begon M, Ward MJ, Fèvre EM (2017) Urbanization and disease emergence: dynamics at the wildlife–livestock–human interface. Trends Ecol Evol 32(1):55–67

ICSU (2017) A guide to sdg interactions: from science to implementation. International Council for Science, Paris, France

Lee BY, Bartsch SM, Mui Y, Haidari LA, Spiker ML, Gittelsohn J (2017) A systems approach to obesity. Nutrition Rev 75(suppl_1): 94–106

Low W-Y, Lee Y-K, Samy AL (2015) Non-communicable diseases in the Asia-Pacific region: prevalence, risk factors and community-based prevention. Int J Occup Med Environ Health 28(1):20–26

McBean G, Cooper R, Joakim E (2017) Coastal cities at risk (CCaR) : building adaptive capacity for managing climate change in coastal megacities

Newell B, Proust K (2018) Escaping the complexity dilemma. In: König A, Ravetz J (eds) Sustainability science: key issues. Earthscan, Routledge, Abingdon, Oxon

Newell B, Siri J (2016) A role for low-order system dynamics models in urban health policy making. Environ Int 95(October):93–97

Ramaswami A, Russell AG, Culligan PJ, Sharma KR, Kumar E (2016) Meta-principles for developing smart, sustainable, and healthy cities. Science 352(6288):940–943

Saksena S, Duong NH, Finucane M, Spencer JH, Tran CC, Fox J (2017) Does unplanned urbanization pose a disease risk in Asia? The case of avian influenza in Vietnam. AsiaPacific Issues. East-West Center, Honolulu, Hawaii

SCHEMA (2018a) Schema case studies: applying systems thinking to urban health and wellbeing. In: Tan D, Siri J (eds). United Nations University International Institute for Global Health, Kuala Lumpur, Malaysia

SCHEMA (2018b) Thrive connect: linkages in everyday life. In: Tan D, Siri J (eds). United Nations University International Institute for Global Health, Kuala Lumpur, Malaysia

Schneider A, Mertes CM, Tatem AJ, Tan B, Sulla-Menashe D, Graves SJ, Patel NN et al (2015) A new urban landscape in East-Southeast Asia, 2000–2010. Environ Res Lett 10(3):34002

Siri JG (2016) Sustainable, healthy cities: making the most of the urban transition. Public Health Rev 37(October):22

Tan DT, Gong Y, Siri JG (2017) The impact of subsidies on the prevalence of climate-sensitive residential buildings in Malaysia. Sustainability 9(12):2300

United Nations Department of Economic and Social Affairs, Population Division (2017) World urbanization prospects: the 2017 revision. United Nations, New York, USA

Wen F, Bedford T, Cobey S (2016). Explaining the geographical origins of seasonal influenza A (H3N2). In: Proceedings of the royal society b: biological sciences, 283(1838)

WHO (2014) Health in all policies: Helsinki Statement, framework for country action : The 8th global conference on health promotion jointly organized by. World Health Organization, Geneva, Switzerland

WHO (2016) Health as the pulse of the new urban agenda: united nations conference on housing and sustainable urban development, Quito, October 2016. World Health Organization, Geneva, Switzerland

Yang J, Siri JG, Remais JV, Cheng Q, Zhang H, Chan KKY, Sun Z et al (2018) The Tsinghua-Lancet commission on healthy cities in China: unlocking the power of cities for a healthy China. Lancet 391(10135):2140–2184

A Systems Approach to Urban Health and Well-being Has Come of Age in Latin America and the Caribbean

Manuel Limonta

Key Messages

1. The Latin American and the Caribbean region have one of the highest urbanization rates in the world.
2. Access to health services, epidemiologic transition and chronic non-communicable diseases, training and distribution of human resources in health, inequalities in health, and financing health systems are among the most important health challenges.
3. El Salvador started in 2010 the implementation of a profound reform of the health system which has led to significant health improvements that has triggered the establishment of the Salvadoran Urban Health Model.
4. The new model of urban health in El Salvador is inspired by the systems approach and has potential to be exemplary for the entire LAC region.
5. ICSU ROLAC has played the role of facilitator in all these developments together with the participants mentioned. Also, our Secretariat has worked with this Urban Health Initiative with the orientation and alignment of our ICSU Health and Wellbeing Program applied to our region.

Latin America and the Caribbean region have one of the fastest urbanization rates in the world. Hundreds of mid-sized cities are emerging. More than 80% of the region's population lives in cities. In Venezuela, Argentina, Uruguay, Brazil, and Chile, more than 90% of the countries' population is urban. Thirty percent of the population of Latin America and the Caribbean does not have access to health care. Seventy-four million people do not have adequate sanitation and less than 20% of wastewaters and sewage is treated, leading to serious health risks (UN Habitat 2012) (Fig. 1).

M. Limonta (✉)
Regional Office for Latin America and the Caribbean, International Council for Science, San Salvador, El Salvador
e-mail: manuel.limonta@icsu-latin-america-caribbean.org

© Zhejiang University Press and Springer Nature Singapore Pte Ltd. 2020
F. W. Gatzweiler (ed.), *Urban Health and Wellbeing Programme*,
Urban Health and Wellbeing, https://doi.org/10.1007/978-981-15-1380-0_3

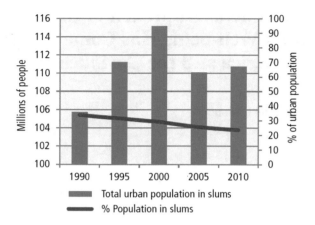

Fig. 1 Urban slum population trends, 1990–2010 UN Habitat 2012

In 2012, an estimated 111 million people lived in slums, and Latin America is home to cities that are among the most unequal and unsafe in the world. Latin American cities are leading in being the most murderous cities on earth, with cities in El Salvador, Honduras, Mexico, and Guatemala leading the charts (UN Habitat 2012). The World Economic Forum on Latin America (2016) estimates that the number of people who are living with a non-communicable disease (NCD) is over 200 million. NCDs are associated with 79% of all deaths. Thirty-five percent of deaths caused by NCDs is premature, i.e., they occur between 30 and 70 years of age.

$21.3 trillion in economic losses in low- and middle-income countries could be caused by NCDs over the next two decades. The five of the most important health challenges include: 1. Access to health services, 2. Epidemiologic transition and chronic non-communicable diseases, 3. Training and distribution of human resources in health, 4. Inequalities in health, and 5. Financing health systems.

At least 100 million people in the region are exposed to air pollution above the limits recommended by the World Health Organization (WHO). Land-based transport is the greatest cause of air pollution in the cities of Latin America and the Caribbean. Others are coal and heavy oil-fueled power stations, and industrial production. Annual average PM10 levels (2000–2004) also exceeded the WHO's recommended maximum in most cases, and the concentrations even exceed the standard set by cities themselves (Greene and Sánchez 2012).

The case of air pollution shows that urban health issues are highly interconnected. Better understanding how they are interconnected and aiming at issues which are systemically linked promises the co-benefits from more efficient and effective policy measures. Promoting a systems approach to urban health in the region has therefore been one of the main goals of the International Council for Science's Regional Office for the region (ICSU ROLAC). Urban health is one of the ICSU ROLAC's priority areas, and the office has been active since its recent move from Mexico to El Salvador.

Fig. 2 On June 19, 2017, the Ministry of Health officially launched the new Urban Health Model in El Salvador. Participants of the high-level meeting included the Chinese Embassy in El Salvador, and the Technical Secretary of Planning of the Presidency Lic. Alberto Enríquez

The ROLAC office moved to El Salvador in August 2016, and in October 2016, a first workshop was conducted with the Ministry of Public Health of El Salvador (MINSAL) and the Ministry of Public Works.

The workshop was attended by 340 participants including distinguished representatives of the Government as ministers, advisers, representatives of the educational system, authorities of the National University, associations involved in environmental and health research, and a significant percentage of college students who were very interested in the subject.

One of the findings of the workshop was the lack of inter-sectoriality. That reason encouraged ICSU ROLAC to begin establishing and coordinating a working group on urban health. The Vice Minister for Health from El Salvador, together with the Vice Minister of Science and Technology, held a meeting with the ROLAC office to encourage the work of different ministries and other organizations coordinated by ICSU ROLAC to create synergies and face the work of Urban Heath in El Salvador.

On March 20, 2017, under the coordination of ICSU ROLAC, the first national Urban Health Working Group meeting was held. It was attended by Vice Ministers, Directors, and Advisors of key institutions in the country, to expose existing programs, gaps, and challenges in the area of urban health and to establish a line of action for the implementation of a New Urban Health Model in El Salvador (Fig. 2).

The Urban Health Working Group is composed of the following institutions:

- Vice Ministry of Science and Technology
- Ministry of Health
- Ministry of Environment and Natural Resources
- Ministry of Public Works
- Ministry of Justice and Public Security
- Vice Ministry of Housing and Urban Development
- COAMSS OPAMSS
- National University of El Salvador
- Director of Transportation

- Ministry of Education
- Secretary of Culture
- National Institute of Sports of El Salvador
- National Institute of Youth
- National University of El Salvador
- Environmental Impact (Media).

The working group has made important progress, such as the consolidation of an official National Matrix with information on the programs currently being carried out by the different ministries in the area of urban health in El Salvador, in order to work on these programs in an integrated way among all ministries.

In the words of the Minister of Health, Dr. Violeta Menjívar, "the main objective of the Urban Health Model is to reduce social exclusion, protect and restore the environment, promote human development and build a healthier city for all." The new model of urban health is being implemented in the country and will have an important impact at the regional level where it can be the starting point for other countries in the region to replicate this example of integrated governance for health pioneered by El Salvador.

During its launch, it was stated that the core of Salvadoran Urban Health Model starts on what has been coined as "Good living" which can be attained by implementing public policies which positively affect the individual, the family, and the community. The model then presents seven dimensions: coexistence, mobility, environment, society and well-being, habitat and infrastructure, governance and integration, and education and culture. Finally, the model encloses its actioning within three main strategies: social participation, inter-sectoriality, and intra-sectoriality.

It is important to highlight that the Salvadoran Urban Health Model was conceived within the context of El Salvador's health reform which began in 2010, restated health as a fundamental human right. The country initiated a profound and innovative reform of the health system, setting in motion a road map toward Universal Access to Health, breaking geographical, economic, and technological barriers to guarantee the right to health of the population, by strengthening the public health system.

Dr. Manuel Limonta, Regional Director of the International Council for Science (ICSU), explained that the Urban Health Model of El Salvador was created on a scientific approach inspired by the systems approach to Urban Health and well-being under ICSU, and based on the previous work done by the MINSAL. He announced that ICSU ROLAC in conjunction with the TWAS World Academy of Sciences an International Urban Health Workshop in September 2017 would organize the Second Urban Health Workshop in El Salvador, which eventually was attended by Young Scientists from poor countries of Latin America and the Caribbean, addressing the theme of urban health. Such workshop was planned by all members of the Urban Heath Working Group.

The Second International Urban Health Workshop was held in San Salvador, El Salvador on September 28–29, 2017. The activity was organized by ICSU ROLAC, the World Academy of Science (TWAS), the Ministry of Health of El Salvador, and ICSU ROLAC's Urban Health Working Group. The workshop aimed to facilitate and encourage the interaction of successful experiences and ties of cooperation among the participants of the event. That workshop also represented a great opportunity to show the Salvadoran model with other countries of the region, which was perceived as pioneering.

The workshop has facilitated the participation and interaction at the national and regional level of the participants, emphasizing the importance of facing the urban health situation from an integral, multidisciplinary, and inter-institutional approach.

On July, it was decided to run a pilot test in San Salvador, since it is the capital of El Salvador, and it accounts for a 27.19% of the country's population that is 1,773,557 inhabitants (DIGESTYC 2016). From the 14 municipalities of San Salvador, seven were selected in order to run a pilot test. It was at this stage, by mid-October, when the Pan American Health Organization introduced an "**Action Tool for Healthy Cities,**" which was presented in different workshops in order to train the trainers of the seven municipalities selected for the pilot test. The objective of the workshop and the tool itself was to provide guidance to take action measures to promote healthy cities, municipalities, and communities; and to do so, it provides questions to consider, key activities, examples, and resources to help with the health promotion communities (Fig. 3).

In El Salvador, the tool is expected to be used to implement the Urban Health Model and its Implementation Plan (MINSAL 2017). The core of the model is to promote what has been labeled as "Good Living" in El Salvador through strategies of social participation, inter-sectorality and intra-sectoriality. "Good Living" embraces the following dimensions: coexistence, mobility, environment, society and well-being, habitat and infrastructure, governance and integration, and education and culture.

Since the Salvadoran Urban Model comprises all the previously mentioned dimensions, each of the seven municipalities chosen for the pilot test selected different issues that directly caused a negative impact within their communities, in order to

Fig. 3 7 Municipailities selected as pilot sites for the Urban Health Model in El Salvador

tackle them and bring about a change of paradigm, through the inter-sectoral and intra-sectoral approach which this model proposes.

The ROLAC Secretariat has provided support to all the workshops, and it has also accompanied them closely. The execution of these workshops within the communities moves the Salvadoran Urban Health Model from theory to practice, engaging the citizens of the cities in which the systems approach is applied.

By means of the Urban Health Model, it has been possible to align the work of the entire country, through the ministries which belong to the working group and also to encourage the interest of governmental actors in the subject. In addition, common citizens and members of the communities have begun to get involved in the program.

The ROLAC has envisioned short, middle, and long-term goals. In the short term, it has been already possible to attain a diagnostic of the urban health state in El Salvador, the creation of a working group, and a national matrix of programs currently carried out by ministries in the area of urban health in El Salvador. In the middle-term goal, it has been possible to have an implementation plan which is in its early execution. In the long term, it is expected to obtain transformational change in urban health. It is worth mentioning that what has been achieved so far has been the result of coordination, management, and integrated governance by taking a systems approach.

Looking into the future ROLAC plans to organize a Central American Workshop in which the model can be introduced to the region. That will be an opportunity to establish links of cooperation and association with the countries of the region, creating the base for a larger initiative with countries from the entire LAC region.

Other global regions begin to show an interest in the Urban Health Model. ROLAC has been invited to present the model at "Future Earth Health Knowledge-Action Network Workshop" in Xiamen, China, on December 2–4, 2017, at the 9th Session of the World Urban Forum in Kuala Lumpur, February 7–13, 2018, and in Guatemala at the Symposium of Urban Health Policies on May 18, 2018. The Chinese Academy of Science has invited representatives of ROLAC and the Urban Health Working Group from El Salvador to present the Urban Health Model at a Belt and Road symposium which will be held October 16–18, 2018, in China.

References

DIGESTYC (2016) Estimaciones y proyecciones de la población. San Salvador, El Salvador. Available at http://www.digestyc.gob.sv/index.php/novedades/avisos/540-el-salvador-estimaciones-y-proyecciones-de-poblacion.html

Greene J, Sánchez S (2012) Air quality in Latin America: an overview. Clean Air Institute, Washington, D.C., USA

Ministerio de Salud de El Salvador–MINSAL (2017) Modelo de Salud Urbana y Plan de Implementación. Available at: http://www.salud.gob.sv/archivos/comunicaciones/archivos_comunicados2017/pdf/Modelo_de_Salud_Urbana_y_Plan_de_Implementacion.pdf

UN Habitat (2012) The state of Latin American and Caribbean cities 2012. Towards a new urban transition Nairobi. United Nations Human Settlements Programme, Kenya

World Economic Forum (2016) These are the 5 health challenges facing Latin America Accessed 20 Aug 2017 at: https://www.weforum.org/agenda/2016/06/these-are-the-5-health-challenges-facing-latin-america/.See also San Salvador Urban Health Model (in Spanish): https://youtu.be/XUzbvvHeYR0

A Systems Approach to Urban Health and Well-being Has Come of Age in Africa

Tolu Oni

Key Messages

1. Health is a central part of development, but the majority of factors that influence health lie outside the health sector.
2. Rapid urbanization across Africa is resulting in changing exposures which influence health, often negatively, but can be harnessed for health creation.
3. To achieve the African Union's Agenda 2063 vision of shared well-being, new approaches to improving health and well-being are needed.
4. Such approaches will require coordination between health and non-health sectors to ensure Healthy Public Policies.
5. Systems approach address the complex Web of interrelated factors that influence health in cities, promote new forms of intersectoral dialogue and data sharing, and provide tool kits to develop intersectoral health promotion plans and interventions

The growing majority of urban dwellers (62%) in Africa now live in informal conditions that, without access to basic services or public amenities, expose residents to greater health risk, and health-care systems, are unable to provide affordable or comprehensive cover. Unplanned and unmanaged growth across urban Africa and high rates of poverty are associated with exposures that increase risk of both infectious and non-communicable diseases (NCDs). For example, the burden of diabetes in Africa is expected to increase by 110% between 2013 and 2035. The complex nature of factors driving these changing patterns of disease necessitates a systems approach.

The Healthy Cities approach is an example of a systems-based approach, which recognizes that enhancements to population health will come about through improvements in environmental, socio-cultural, and economic conditions, coupled with

T. Oni (✉)
Clinical Senior Research Associate, MRC Epidemiology Unit, University of Cambridge, Cambridge, UK
e-mail: tolullah.oni@uct.ac.za

© Zhejiang University Press and Springer Nature Singapore Pte Ltd. 2020
F. W. Gatzweiler (ed.), *Urban Health and Wellbeing Programme*,
Urban Health and Wellbeing, https://doi.org/10.1007/978-981-15-1380-0_4

behaviour changes. Whilst individual behaviour change has been the focus of health promotion strategies in the past, and it is important to note that these urban exposures can undermine the ability to change behaviour. A central philosophy of this approach is that health should be seen as a core part of overall development.

In 2015, Agenda 2063 was adopted by the African Union (AU) as the united vision for development in Africa. Health is a core part of overall development, and cannot be achieved without a concerted effort. To ensure interventions in the economic, environmental, and social spaces do not negatively impact health. Beyond merely limiting adverse health impacts, there is an opportunity to embrace the responsibility to create and support healthy public policies across all sectors. The Healthy Cities approach is not new in African cities.

In 1990s, several African cities adopted this systems approach, from Accra and Bangui, to Cape Town and Dar es Salaam. While strategies differed, a core component to this was the need to develop a city health profile: not only the health status of the city but the processes that undermine health. This profile is then utilized to develop a city health plan, a strategic planning document developed in consultation with a wide range of stakeholders across sectors. Given the significant increase in urbanization in the last 30 years, these intersectoral approaches are needed now more than ever.

The Research Initiative for Cities Health and Equity (RICHE) was established to bring together researchers from multiple fields, practitioners and policymakers to co-create and evaluate intersectoral interventions to improve health outcomes, resiliencies and address vulnerabilities introduced across the stages of life in cities across Africa.

In 2015, a RICHE workshop organized in Cape Town by Prof Oni, brought together researchers, policymakers and practitioners with a wide range of experience in urban health in Africa. A subsequent symposium of urban health researchers, practitioners and policymakers was organized in 2017 to identify opportunities for research collaboration across Africa, to explore policy priorities and needs for healthy cities, and to further ideas on the development of a training curriculum on urban health for Africa.

These workshops identified key focus areas to advance urban health in Africa:

A. Obesity and food insecurity: The urban food environment in Africa has changed rapidly with increased access to, and consumption of, sugar, salt, and processed foods. This dietary change has been associated with increasing rates of obesity, diabetes, and cardiovascular disease. Given the high prevalence of food insecurity occurring alongside rising rates of obesity, higher in women, the need was identified for interventions that simultaneously engage both food insecurity and obesogenic food environments, and that engage with the informal food economy. There is evidence that pre-conception and peri-partum nutrition is vital for fetal health and the subsequent health of the child into adulthood. Therefore, nutritional interventions to reduce obesity and improve food security are required, with a particular focus on women-headed households, the first 1000 days and female adolescents.

> … despite pressing needs driven by Africa's considerable and complex burden of disease and high levels of health inequity, urban health and urban health equity have not yet emerged as major research and policy priorities in Africa, and as such South Africa, like many other African countries, lags behind in addressing these issues. Oni et al. (2016)

B. Healthy urban environments: Poor housing quality and inadequate urban planning have been linked to both infectious and NCDs, from pneumonia and diarrhoea, to asthma and obesity. A systems approach considers item 72b of Agenda 2063s call to action which calls for ensuring Africans have access to decent affordable housing. This approach moves beyond the physical structure of the house and access to health-care services, and asks how to ensure such housing opportunities can promote health and well-being, by ensuring a settlement that creates health, through access to healthy food, adequate waste removal, sanitation, ventilation, and opportunities for safe physical activity. To this end, ongoing research in Cape Town is exploring barriers and facilitators of incorporating health objectives into housing policy, with a view to generating evidence to support co-creation of healthy intersectoral human settlements interventions.

C. Urban health governance and policy: This systems approach requires coordination across health and non-health sectors, and should be shaped by individual and collective identity, the lived environment, and urban policies. The shortage of intersectoral implementing groups to promote accountability for population health across sectors and levels of government was identified as a key gap to develop and implement an intersectoral city health plan. This highlights the importance of building capacity of local governments to design and implement such a plan.

D. Community strengthening for healthy cities: Citizen engagement and participatory processes, which emphasise the lived experiences of residents, including their capabilities, preferences, and needs, in all stages of policy development

Fig. 1 Changing face of
cities in Africa

and implementation are important to address urban determinants of health. A
substantial barrier to citizen engagement is poor education about processes of
urban governance to improve health, and this would need to be addressed in any
City Health Strategic Plan.
E. Migration, urbanization, and health: Agenda 2063 recognizes movement of peo-
    ple as a fundamental characteristic of the African continent, and aspires for an
    Africa with free movement of people. A systems approach to urban health recog-
    nizes the potential for migration and mobility to adversely impact on health, and
    considers the opportunity for healthy migration and mobility through a better
    understanding of patterns of circular migration and differential health exposures
    at each stage of migration; as well as the ways in which health, economic, and
    social policies and systems can best respond to mobility.

To conclude, the aspirations of Agenda 2063 reflect the desire for shared prosperity
and well-being, for unity and integration, for a continent of free citizens and expanded
horizons, where the full potential of women and youth, boys and girls are realized,
and with freedom from fear, disease and want.

The first aspiration of Agenda 2063, for a prosperous Africa based on inclusive
growth and sustainable development, where African countries are amongst the best
performers in global quality of life measures, recognises the need for African people
to have a high standard of living, and quality of life, sound health and well-being.
This aspiration further recognizes cities as hubs of cultural and economic activities,
where people have access to affordable and decent housing including housing finance
and all the basic necessities of life such as water, sanitation, energy, public transport,
and ICT.

Beyond merely the vision of skyscrapers and shacks (Fig. 1), a systems approach
to urban health exposes the hidden face of the city, and reveals the essence of the
city's characteristics that influence health, as captured by the 8 S's of urban exposure
(Fig. 2): Sugar and salt (the food environment); Safe housing and social cohesion;
Smoke (indoor and outdoor air pollution) and smoking; Sleep and stress; Sports and
recreation; Sanitation and water; Substance and alcohol abuse; and (unsafe) Sex. A

ugar and salt
afe housing and social cohesion
moke and smoking
leep and stress
ports and recreation
anitation and water
ubstance and alcohol abuse
ex

Fig. 2 Hidden face of the city. 8 S's of urban exposure that influences health

systems approach recognises that these Agenda 2063 aspirations are interconnected; that sound health and well-being cannot be achieved without due consideration of, and a focus on intersectoral strategies that harness desirable city characteristics for health. Improving performance on quality of life measures cannot be achieved without a focus on the most vulnerable in the population. Therefore, implementing such strategies must take on an equity lens, to ensure those most vulnerable are reached.

Urban exposures (as represented by the 8 S's) represent a complex Web of interrelated factors, with often-competing interests and incentives. Beyond merely addressing these exposures, harnessing these for health and well-being will necessitate a systems approach that promotes new forms of intersectoral dialogue and data sharing, and that critically engages with the ensuing trade-offs and benefits. This calls for a re-visiting of the Healthy Cities approach, learning from lessons of the past, to develop intersectoral health promotion plans for cities across Africa.

The systems approach can further support development of toolkits to facilitate this process, as well as mechanisms that foster and share interstitial science which ensures that evidence generated addresses these inter-linkages and facilitates processes that support intersectoral science/policy interaction at the community, city, national, and regional levels.

Reference

Oni T et al (2016) Urban health research in Africa: themes and priority research questions. J Urban Health Bull New York Acad Med 93(4). https://doi.org/10.1007/s11524-016-0050-0

Facilitating the Governance of Urban Sustainability and Resilience Transitions with Knowledge-Action Systems Analysis

Tischa A. Muñoz-Erickson

Key Messages

1. Transition governance models are needed to help cities address sustainability and resilience challenges through collaborative, integrative, and multi-actor network approaches.
2. Governance innovations require that we understand where are we starting from (e.g., how existing institutional conditions work) and where are potential leverage points for innovations and change in governance.
3. KASA is a systems-based framework to help map current governance networks, evaluate the extent that the structure, social preferences, and knowledge systems of these networks enable or constrain transitions, and identify leverage points or interventions for change.
4. The application of KASA in San Juan, Puerto Rico, revealed that a diverse network of organizations existed, including civic organizations, but there were sites in the network that could pose barriers to transitions, and therefore need institutional innovation.
5. Beyond merely analyzing governance structures, the KASA approach allows an examination of how cities think—what different governance actors know about the city, how they know and experience the city, and how they envision the city.

T. A. Muñoz-Erickson (✉)
Research Social Scientist, International Institute of Tropical Forestry (IITF), Río Piedras, Puerto Rico, USA
e-mail: tischa.a.munoz-erickson@usda.gov

© Zhejiang University Press and Springer Nature Singapore Pte Ltd. 2020   29
F. W. Gatzweiler (ed.), *Urban Health and Wellbeing Programme*,
Urban Health and Wellbeing, https://doi.org/10.1007/978-981-15-1380-0_5

1 Wicked Urban Resilience Challenges

On a global level, cities are increasingly leading the way in developing actions to address sustainability and resilience challenges. Yet, many of these challenges, such as climate change, public health, and social justice, are too large, dynamic, and complex for city governments to address on their own. Governance and policy scholars call for a shift from top-down managerial government model to more collaborative, multi-actor network approach. This new transition governance model is characterized by:

1. systems-based and flexible management approaches that do away with agency boundaries in favor of institutional integration and coordination,
2. 'opening' of governments structures to include multiple voices, values, and visions in the development and steering of transition pathways,
3. co-production of knowledge, scenarios, and strategies where government officials, civil society organizations, private sector, and researchers collectively identify problems, produce knowledge, and put that knowledge into action through collaboration, synergy in implementation, and adapting processes.

These governance innovations require that we understand where are we starting from (e.g., how existing institutional conditions work) and where are potential leverage points for innovations and change in governance. This project seeks to support the research, design, and practice of urban governance transitions through the development of a systems-based governance analysis framework—the knowledge-action systems analysis (KASA). KASA is an interdisciplinary framework to help map current governance conditions and networks relevant to sustainability and resilience, evaluate the extent that the structure, social preferences, and knowledge systems of these networks enable or constrain transitions, and identify leverage points or interventions for change (Muñoz-Erickson 2014). Using tools and approaches from institutional, social networks, and knowledge systems analysis, the KASA aids city actors in knowing the 'terrain' of sustainability and resilience actors and initiatives in their cities, and to identify ways to better connect and work together in building climate resilience. Specifically, KASA provides a diagnosis of governance by,

- Describing who are the key actors involved (and not involved) in urban governance, and assess their perceptions, visions, and preferred actions with which to address sustainability and resilience,
- Exploring opportunities for improving connections, knowledge sharing, and collaboration among the multiple actors involved in sustainability and resilience efforts,
- Providing recommendations on key sources of knowledge and capacities needed to anticipate the future uncertainties and envision potential strategies that bring both resilience and sustainability,
- Support the co-creation of the future scenarios with multiple practitioners and stakeholders in cities to develop visions, goals, and strategies for urban sustainability and resilience transitions.

2 Illustration of the KASA Approach: Urban Land Sustainability in San Juan, Puerto Rico

In 2009, increasing development and conversion of green space to construction and cement in the city of San Juan, Puerto Rico, especially in coastal areas and the forested headwaters of the city's main watershed, was producing numerous flood hazards throughout the city. Despite having a municipal land-use regulatory framework that included protection of these green areas as part of the sustainable development of the city, unsustainable land development practices were still taking place.

The application of KASA in San Juan involved the mapping and analysis of the organizations and networks relevant to land-use planning and sustainability, the frames and knowledge that were circulating across the network, and the influence (or power) that actors had on how that knowledge was applied in the land-use governance context. While the analysis revealed that a diverse network of organizations existed, including civic organizations, that where involved in the production and use of knowledge regarding land use (Fig. 1a), there were sites in the network that could pose barriers to the design of urban sustainability and resilience transitions, and therefore need institutional innovation. These included, for instance,

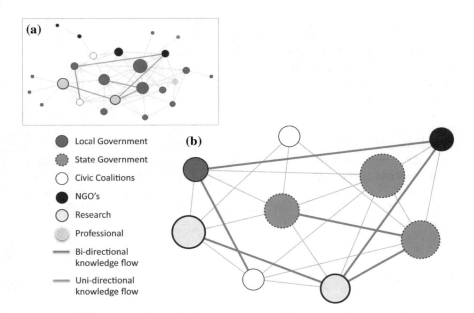

Fig. 1 Network of knowledge flow among organizations involved in land-use governance in San Juan, Puerto Rico. The figure on the top left **a** shows all organizations linked through knowledge flows. Different weights of the nodes, means different levels of centrality, with greater nodes having greater influence over knowledge flow. The larger figure on the right **b** shows only the central actors in the network that have higher degree centrality and betweenness (i.e., brokers) and the reciprocal ties among them (in orange). Muñoz-Erickson and Cutts (2016)

(1) a significant breakdown in knowledge flow between the Municipality and the state's planning agency that acted as a barrier in communicating knowledge of local conditions to the state agency (Fig. 1b),
(2) distinct power asymmetries between the Municipality's visions and knowledge systems which included social dimensions of urban planning (e.g., quality of life and equity goals) and the state's hegemonic ideas of the city as a node for regional economic power,
(3) fragmentation in the knowledge systems tasks and functions of organizations relevant to land-use planning and decision-making instead of collaboration and alignment of agendas and strategies,
(4) knowledge asymmetries were observed, with conventional knowledge types associated with state administration, such as economic and technocratic approaches to planning, have more influence in the network over other alternative types of knowledge (e.g., local, political, social, etc.).

3 Future Directions for KASA

Beyond merely analyzing governance structures, the KASA approach allows an examination of how cities think—what different governance actors know about the city, how do they know and experience the city, how they envision the city. Cities are more than the physical and institutional infrastructures that service an urban population; they are also spaces where a high diversity of actors and their knowledge systems come together in networks that catalyze new ideas and innovations. The Urban Resilience to Extreme Events Sustainability Research Network (UREx SRN), an international network of researchers and practitioners addressing urban resilience and sustainability challenges through actionable knowledge in ten cities in the USA and Latin America, is now applying the KASA approach to better understand the urban resilience governance context, how different actors are envisioning and innovating on resilient urban futures and scenarios, and identify potential interventions for innovations and change in knowledge and governance systems for resilience and sustainability transitions.

References

Muñoz-Erickson TA (2014) Co-production of knowledge-action systems in urban sustainable governance: the KASA approach. Environ Sci Policy 37:182–191
Muñoz-Erickson TA, Cutts B (2016) Structural dimensions of knowledge-action networks for sustainability. Curr Opin Environ Sustain 18:56–64

Advancing Urban Health and Wellbeing Through Collective and Artificial Intelligence: A Systems Approach 3.0

Franz W. Gatzweiler

Key Messages

1. Complex problems of urban health and wellbeing cause millions of premature deaths annually and are beyond the reach of individual problem-solving capabilities.
2. Collective and artificial intelligence (CI + AI) working together can address the complex challenges of urban health.
3. The systems approach (SA) is an adaptive, intelligent and intelligence-creating, "data-metabolic" mechanism for solving such complex challenges.
4. Design principles have been identified to successfully create CI and AI. Data-metabolic costs are the limiting factor.
5. A call for collaborative action to build an "urban brain" by means of next-generation systems approaches is required to save lives in the face of failure to tackle complex urban health challenges.

1 Challenges of Urban Health and Wellbeing

> As our world becomes more and more closely connected, through all kinds of electronic communication, it will become more and more useful to view all the people and computers on our planet as part of a single global brain. And perhaps our future as

F. W. Gatzweiler (✉)
Programme on Urban Health and Wellbeing, the Global Interdisciplinary Science Programme on Urban Health and Wellbeing, Xiamen, China
e-mail: gatzweiler02@gmail.com

© Zhejiang University Press and Springer Nature Singapore Pte Ltd. 2020
F. W. Gatzweiler (ed.), *Urban Health and Wellbeing Programme*,
Urban Health and Wellbeing, https://doi.org/10.1007/978-981-15-1380-0_6

a species will depend on how well we're able to use our global collective intelligence
to make choices that are not just smart but also wise. Malone (2015)

Many problems of urban health and wellbeing, such as pollution, obesity, aging, mental health, cardiovascular diseases, infectious diseases, inequality, and poverty (WHO 2016), are highly complex and beyond the reach of individual problem-solving capabilities. Biodiversity loss, climate change, and urban health problems emerge at aggregate scales and are unpredictable. They are the consequence of complex inter-actions between many individual agents and their environments across urban sectors and scales. Another challenge of complex urban health problems is the knowledge approach we apply to understand and solve them. We are challenged to create a new, innovative knowledge approach to understand and solve the problems of urban health. The positivist approach of separating cause from effect, or observer from observed, is insufficient when human agents are both part of the problem and the solution.

Problems emerging from complexity can only be solved collectively by applying rules which govern complexity. For example, the law of requisite variety (Ashby 1960) tells us that we need as much variety in our problem-solving toolbox as there are different types of problems to be solved, and we need to address these problems at the respective scale. No individual has the intelligence to solve emergent problems of urban health alone.

2   Collective and Artificial Intelligence

Collective intelligence (CI) is defined as the intelligence of a group created by sharing knowledge and working together toward the same end to solve common problems. This is essential for addressing the challenges that emerge from complexity and uncertainty since it exists beyond the reach of individual problem-solving. Design principles for CI have been referred to as "genes" (Malone et al. 2010). These "ge-nes" are the principles upon which we can find answers to the questions of what needs to be done, why and how, and by whom. For example, the genes that drive motivation, e.g., love and glory, are principles that successfully operated in the col-lective development of the Linux operating system. Ostrom (2005: 258) referred to them as "design principles." Wikipedia, TEDx, or the Urban Health Collaborative at Drexel University and the Climate CoLab at Michigan Institute of Technology, are additional examples of CI.

We now know that people are more successful at solving common problems collectively than in hierarchies, states, or markets. Interaction, communication, and perception of commonly-faced problems are essential prerequisites. Smaller group size and homogeneity tend to be supportive, yet, broad participation in decision-making and strong interpersonal skills, as well as female presence and diversity in group composition (Malone and Klein 2007), are more closely associated with CI. Speaking a common language and sharing mental models and conceptions (Dyball and Newell 2015) to define the challenges and understand how they are interrelated lead to better solutions. Success of individual or expert intelligence depends solely on context when either first-time decisions are made or decision-makers learn from past mistakes. Katsikopoulos and King (2010) found that groups were generally more intelligent decision-makers, whereas individuals were only successful when relying on previous decision outcomes. Thus, in order to arrive at a different paradigm, by definition, we must look to CI.

Artificial intelligence (AI) has emerged in an attempt to improve knowledge creation by processing big data with high-performance computing and machine learning. As a result of the evolution of intelligence, we have reduced the costs of data processing and learning. AI makes transforming data into knowledge faster and more cost effective. Large groups of people can better communicate and act collectively if facilitated by AI. Multiple examples now exist to show us how CI and AI, operating together (CI + AI), enable people to understand and address complex problems of urban and planetary health on a global scale (Weld et al. 2014).

Essential for creating CI and AI are the costs of exchanging and processing information. Data processing, interaction, and collective action are factors that pose the greatest obstacle to building CI + AI because the costs are essentially determined by group size. As the group increases, the more unmanageable and costly become information exchanges, data flows, processing of data, and coordination of collective action.

In the past, the evolution of hierarchy (and modularity within hierarchy) developed out of the need to problem solve on a large social scale. It overcame the same obstacle of decreasing marginal intelligence with increasing group size (Powers and Lehmann 2017) when population size increased by becoming an important driver of network performance and adaptability (Mengistu et al. 2016).

In today's rapidly changing environments and growing societies, hierarchies continue to face increasing costs of information processing. As a result, they have turned to heterarchical-participative networks to solve problems characterized as complex, uncertain, and dynamic. Similar to polycentric organization (Aligica and Tarko 2012), the heterarchical organization has multiple command centers to accommodate the increasing participative cultures, autonomy, and high levels of self-determination, as well as ongoing dialogue—a structure more conducive for building CI. "The power of heterarchies lies in their flexibility and capacity for innovation" (Schwaninger 2006: 31).

3 Evolution of the System Approach

The systems approach (SA) to urban health and wellbeing (Gatzweiler et al. 2017; Bai et al. 2016) has been developed to better understand and solve complex problems of urban health. It combines both systems methods and models to understand problems emerging from complexity and the engagement of stakeholders. Participatory modeling of complex urban health problems, like resilience, has been successfully developed by the Ecological Sequestration Trust. In principle, the SA is an adaptive, recursive, cognitive mechanism at work in nested systems of various social, ecological, territorial, technological, or cyber spaces. The SA is a co-evolutionary process of, let's call it, a "data metabolism," transforming data into knowledge and knowledge into procedural and structural change (referred to as "action" among human agents) in response to external system changes. The mechanism of the SA is a driver of evolution.

The SA operates, for example, in biology, knowledge generation, or in social organization, and action. It can be seen as analogous to the process of the adaptive cycle described by the concept of panarchy (Gunderson and Holling 2002)—a mechanism that is driven by two mutually reinforcing processes: a catabolic process which breaks down complex wholes into smaller parts; and an anabolic process which rebuilds. The process at work can be portrayed as intelligent and as self-creating intelligence. It facilitates the building of collective and artificial intelligence and, in turn, requires both types of intelligence for data processing, knowledge creation, and action (Komninos 2008).

The SA itself is evolving. Its evolution (Fig. 1) is characterized by data collection and processing mechanisms. These processes are most likely to happen where and when the costs of information exchange and flow are lowest. The lower the costs, the lower the resistance to the data-metabolic process and the better the SA performs, thus, the SA not only drives the evolution of systems, it evolves. The path of its evolution leads along avenues of least resistance, similar to the way electric voltage in lightning follows a path through the atmosphere.

In the evolution of the SA, it is the SA1.0 that tackles problems of complexity in socio-ecological-technological systems (SETS) as it combines complexity modeling with stakeholder engagement. The process can be very costly, particularly when bringing together engaged experts and citizens, creating a common understanding of the problem, and communicating common goals and outcomes.

Fig. 1 Evolution of the systems approach from SA1.0 to SA3.0

SA2.0 then emerges from any other alternative approach as the most cost effective with regard to its data-metabolism. It integrates engaged citizens into the model which simulates the environment in which urban health problems emerge. What was a "virtual model' in SA1.0 now enters the real world where the engaged citizen plays the game. By playing repeatedly, the human agent learns how to solve complex problems better than before. Individual and collective intelligence can improve and new skills develop. For example, city developers can learn by playing games like Sim City, City Skylines or IBM's City One, or nation-state managers can learn by playing "Ecopolicy."

4 A Callforaction: SA 3.0 and the Urban Brain

Today, the best players of the Chinese board game GO have lost against Google's Deep mind machine using intuition, something once believed to be a uniquely human trait. AI developed by Elon Musk's Open AI team recently won against a world champion of the online multiplayer game DOTA2 showing us that machines can outperform humans. Humans are not simply teaching machines, they are learning from them.

SA3.0 is the next phase in the evolution of the SA. It would advance the human learning process even further by applying AI. In the SA3.0, CI and AI, mutually supporting each other, would produce better solutions in complex problem-solving processes for solving problems of urban health and wellbeing. The algorithms of AI would improve the data processing and learning process, while the rules and regulations in society would support the building of CI. The SA3.0 would be the mechanism at work to drive the mutual enhancement CI + AI to shape the collective brain of a city:

$$(CI + AI) \times SA3.0 = \text{collective urban brain}$$

Working together with CI and AI innovators would advance the urban brain's learning curve by applying variations of the SA in order to build urban intelligence—intelligent cities that are resilient and able to adapt to change for the health and wellbeing of its inhabitants. Facing millions of premature deaths each year necessitates a clarion call for collaborative action to apply SA3.0 for building the collective urban brain in order to overcome the increasing risks of urban health and wellbeing that exists in our urban world today.

References

Aligica P, Tarko V (2012) Polycentricity: from polanyi to ostrom, and beyond. Governance 25(2):237–262

Ashby RW (1960) Design for a brain. the origin of adaptive behavior. Chapman and Hall, London

Bai X, Surveyer A, Elmqvist T, Gatzweiler FW, Güneralp B, Parnell S, Prieur-Richard A-H, Shrivastava P, Siri JG, Stafford-Smith M, Toussaint J-P, Webb R (2016) Defining and advancing a systems approach for sustainable cities. Curr Opinion Enviro Sustain 2016(23):69–78. https://doi.org/10.1016/j.cosust.2016.11.010

Dyball R, Newell B (2015) Understanding human ecology. a systems approach to sustainability. Routledge, London and New York

Gatzweiler FW et al (2017) Advancing health and wellbeing in the changing urban environment. In: Implementing a systems approach. Zhejiang University Press, Hangzhou and Springer, Singapore

Gunderson L, Holling CS (2002) Panarchy: understanding transformations in systems of humans and nature. Island Press, Washington, Covelo, London

Katsikopoulos KV, King AJ (2010) Swarm intelligence in animal groups: when can a collective out-perform an expert? PLoS ONE 5(11):e15505. https://doi.org/10.1371/journal.pone.0015505

Komninos N (2008) Intelligent cities and globalisation of innovation networks. Routledge, London, New York

Malone TW, Klein M (2007) Harnessing collective intelligence to address global climate change innovations: technology. Governance Globalization 2(3):15–26

Malone TW, Laubacher R, Dellarocas C (2010) The collective intelligence genome, MIT sloan management review. Cambridge 51(3):21–31

Malone TW (2015) Building better organizations with collective intelligence: Webinar MIT. Accessed Aug 14, 2017 at: https://www.youtube.com/watch?v=hu4ZXr40bSA

Mengistu H, Huizinga J, Mouret JB, Clune J (2016) The evolutionary origins of hierarchy. PLoS Comput Biol 12(6):e1004829. https://doi.org/10.1371/journal.pcbi.1004829

Ostrom E (2005) Understanding institutional diversity. Princeton University Press, UK and New Jersey

Powers ST, Lehmann L (2017) When is bigger better? The effects of group size on the evolution of helping behaviours. Biol Rev 92:902–920. https://doi.org/10.1111/brv.12260

Schwaninger M (2006) Intelligent organizations: powerful models for systemic management. Springer, Berlin p 31

Weld DS, Lin CH, Mausam, Bragg J (2014) Artificial intelligence and collective intelligence. Accessed Aug 14, 2017 at: https://homes.cs.washington.edu/~weld/papers/ci-chapter2014.pdf

WHO (2016) Global report on urban health: equitable, healthier cities for sustainable development. World Health Organization, Geneva

The Salud Urbana en América Latina (SALURBAL) Project: Learning from Latin America's Cities for a Healthier Future

The SALURBAL Team

Key Messages

1. Latin America is one of the most urbanized regions of the world and includes many cities of varying size with diverse economic, social, and physical environments.
2. The Latin American region is also home to large social inequalities that manifest themselves in the form of significant health inequities within and across cities.
3. The region has experimented with a range of innovative policy initiatives in the areas of urban development, transportation, social inclusion, and the promotion of healthier behaviors, but health impacts have been rarely quantified.
4. The Salud Urbana en América Latina (SALURBAL) project is the first Latin American project to systematically investigate the factors associated with better health and lower health inequities in cities of Latin America.
5. A novel aspect of the project is employing methods to study linkages between urban health to environmental sustainability and using a combination of approaches (observation, policy evaluation and natural experiments, and systems approaches) to evaluate health impacts and identify promising urban policies.
6. Over a five-year period, SALURBAL will engage with scientists, policymakers, and other sectors of civil society to identify key research questions and disseminate findings.

Latin America is the world's most urbanized region; 80% of its inhabitants are concentrated in cities with a projected jump to 90% by 2050 (United Nations 2014). As in other cities all over the world, health in Latin American cities emerges from the complex interplay of social, economic, and physical environments. Diverse populations, large social inequalities, and spatial segregation create and reinforce significant health inequities. Environmental sustainability and population health are

The SALURBAL Team (✉)
Salud Urbana En América Latina (SALURBAL) Team, Philadelphia, USA
e-mail: uhc@drexel.edu

© Zhejiang University Press and Springer Nature Singapore Pte Ltd. 2020   39
F. W. Gatzweiler (ed.), *Urban Health and Wellbeing Programme*,
Urban Health and Wellbeing, https://doi.org/10.1007/978-981-15-1380-0_7

Fig. 1 Informal settlements or *favelas* in Rio de Janeiro, Brazil

mutually reinforcing; factors conducive to sustainable environments promote health, and healthier behaviors have beneficial environmental consequences.

The region faces health risks tied to social inequality, chronic diseases and aging, emerging infectious diseases, violence, and injuries (Briceño-León 2005; Mutaner et al. 2012; Vilalta et al. 2016). Approximately, 1 in 5 urban inhabitants in Latin America lives in a slum or informal settlement (World Bank 2017; UN-Habitat 2012). Latin America has the highest homicide rate in the world, averaging 24 victims per 100,000 inhabitants (UNODC 2013). The region also has one of the highest rates of death and disability due to traffic accidents (Haagsma et al. 2015) (Fig. 1).

Heart disease, cancers, diabetes, and respiratory diseases remain the leading causes of premature death in the region, responsible for 81% of all deaths (PAHO 2014). Physical inactivity (Arango et al. 2013; Gomez et al. 2015), the growing consumption of processed foods and obesity (Fishberg et al. 2016), and persistent malnutrition (World Bank 2005) co-exist in Latin American cities. The rapid growth of many cities has been accompanied by limited planning processes, growing automobile traffic, and poor air quality (Fajersztajn et al. 2017).

An especially novel aspect of SALURBAL is the use of systems approaches to yield insight into how a range of dynamic processes jointly affect health and environmental sustainability. The project identified transportation and food environments as two areas for systems modeling based on (1) the dynamic relations involved; (2) policy interest in the region; and (3) relevance to sustainable development goals.

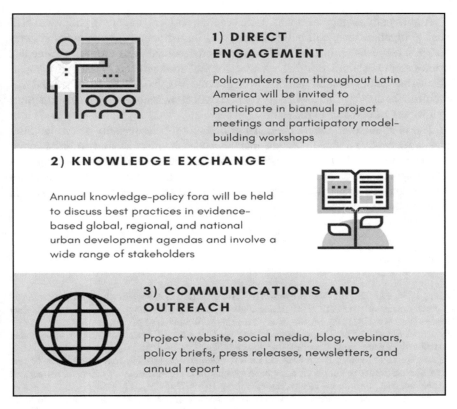

Fig. 2 SALURBAL's planned policy translation activities

Simulation models will allow us not only to identify a plausible range of effects, but also to characterize the conditions under which these effects manifest and identify any unintended consequences. This is of particular relevance to other regions in the world.

A series of participatory group model-building workshops will allow key stakeholders to conceptually map the multiple pathways through which a city's transportation options and food policies impact health. The conceptual maps and causal loop diagrams generated in these meetings will be used to inform the questions and structures that will be employed in formal simulation models in the second phase of the project. SALURBAL's research findings will carry a range of policy implications of relevance to government officials, public health practitioners, urban planners, social and economic development organizations, and the public.

The project will translate research findings into actionable knowledge and engage policymakers and other stakeholders in academia and civil society through direct engagement, knowledge exchange, and communications and outreach at all stages of the project (see Fig. 2).

Additionally, engagement with non-academic partners, including policy and civil society organizations, will build capacity and infrastructure for continued systems research and evaluations. SALURBAL is unprecedented in its cross-discipline and cross-country network, offering a schema for the future projects to promote comparative urban health research. Similarly, large-scale data that will be compiled and harmonized for the first time under SALURBAL will allow innovative policy evaluations in a variety of domains in the future.

The magnitude of Latin America's urbanization, heterogeneity across its cities and an innovative policy landscape make it well-suited for the study of urban environments and health. Thus, SALURBAL will generate generalizable knowledge that can be applied to policies both inside and outside of Latin America to improve population health, increase health equity and promote environmental sustainability in cities all over the world.

References

Arango CM, Páez DC, Reis RS, Brownson RC, Parra DC (2013) Association between the perceived environment and physical activity among adults in Latin America: a systematic review. Int J Behav Nutr Phys Act 10(1):122. https://doi.org/10.1186/1479-5868-10-122

Briceño-León R (2005) Urban violence and public health in Latin America: a sociological explanatory framework. Cadernos de Saúde Pública 21:1629–1648

Fajersztajn L, Saldiva P, Pereira LA, Leite VF, Buehler AM (2017) Short-term effects of fine particulate matter pollution on daily health events in Latin America: a systematic review and meta-analysis. Int J Public Health. https://doi.org/10.1007/s00038-017-0960-y

Fisberg M, Kovalskys I, Gómez G, Rigotti A, Cortés LY, Herrera-Cuenca M et al (2016) Latin American Study of Nutrition and Health (ELANS): rationale and study design. BMC Public Health 16(1):93. https://doi.org/10.1186/s12889-016-2765-y

Gomez LF, Sarmiento R, Ordoñez MF, Pardo CF, de Sá TH, Mallarino CH et al (2015) Urban environment interventions linked to the promotion of physical activity. A mixed methods study applied to the urban context of Latin America. Soc Sci Med 131:18–30. https://doi.org/10.1016/j.socscimed.2015.02.042

Haagsma JA, Graetz N, Bolliger I, Naghavi M, Higashi H, Mullany EC et al (2015) The global burden of injury: incidence, mortality, disability-adjusted life years and time trends from the Global Burden of Disease study 2013. Injury Prevention

Muntaner C, Rocha KB, Borrell C, Vallebuona C, Ibáñez C, Benach J, Sollar O (2012) Clase social y salud en América Latina. Revista Panamericana de Salud Pública 31:166–175

PAHO (2014) Plan of action for the prevention and control of nonocmmunicable diseaes in the Americas. 2013–2019. Retrieved from http://www.paho.org/hq/index.php?option=com_docman&task=doc_view&Itemid=270&gid=27517&lang=en

UN-Habitat (2012) The State of Latin American and Caribbean Cities 2012. Towards a new urban transition. Retrieved from https://unhabitat.org/?mbt_book=state-of-latin-american-and-caribbean-cities-2

United Nations (2014) World Urbanization Prospects: The 2014 Revision Retrieved from https://esa.un.org/unpd/wup/publications/files/wup2014-highlights.Pdf

UNODC (2013) Global Study on Homicide 2013. Trends, Contexts. Data. Retrieved from https://www.unodc.org/documents/gsh/pdfs/2014_GLOBAL_HOMICIDE_BOOK_web.pdf

Vilalta CJ, Castillo JG, Torres JA (2016) Violent crime in Latin American Cities. Retrieved from https://publications.iadb.org/handle/11319/7821

World Bank (2005) The urban poor in Latin America. Retrieved from http://siteresources.worldbank. org/INTLACREGTOPURBDEV/Home/20843636/UrbanPoorinLA.pdf
World Bank (2017) Population living in slums (% of urban population). Retrieved from http://data. worldbank.org/indicator/EN.POP.SLUM.UR.ZS

Antimicrobial Resistance is a Health Risk in Chinese Cities—Now it Has Been Mapped

Yong-Guan Zhu

Key Messages

1. Antimicrobial resistance (AMR) is becoming a serious urban health risk in China.
2. Unlike conventional chemical contaminants, antimicrobial resistant genes can be amplified in the environment through the spread and proliferation of bacteria and across organisms (horizontal gene transfer).
3. Urban wastewater treatment plants (WWTPs) are hotspots of AMR fueled by human consumption of antibiotics and discharge of other toxic chemicals.
4. This calls for urgent action against the dissemination of AMR through rapid urbanisation.
5. A system's approach to reduce the risk of AMR includes the development of surveillance capacities, pubic awareness and training, multi-sector action plans, and international cooperation.

Antimicrobial resistance (AMR) is the property of microorganisms (e.g., bacteria, viruses, and parasites) developed under frequent exposure to antimicrobial drugs, like antibiotics, to resist them and become so-called "superbugs." "Overuse and misuse of antimicrobial medicines accelerate the emergence of resistant microorganisms" (WHO 2015). AMR is threatening human health worldwide. The widespread use of antibiotics in humans and animals is the main selective driving force of the emergence and dissemination of AMR. Antibiotic resistance is a serious form of AMR. Antibiotic-resistant pathogens now occur at high frequencies in clinical contexts. In particular, the frequent presence of multi-antibiotic-resistant "superbugs" in human feces could lead to a return to the pre-antibiotic era. If the trend continues, a growing number of infections can no longer be treated using the current arsenal of drugs.

Y.-G. Zhu (✉)
Chinese Academy of Sciences, Institute of Urban Environment, Xiamen, China
e-mail: ygzhu@iue.ac.cn

© Zhejiang University Press and Springer Nature Singapore Pte Ltd. 2020   45
F. W. Gatzweiler (ed.), *Urban Health and Wellbeing Programme*,
Urban Health and Wellbeing, https://doi.org/10.1007/978-981-15-1380-0_8

Municipal wastewater treatment plants (WWTPs) receive and digest millions of tons of domestic sewage every day. Adults harbor significant quantities of resistant genes in their gut microbiome, and consequently WWTPs, especially untreated influents are likely to be a critical hub for the evolution and spread of from humans derived resistant genes into natural environments.

In China, more than 3700 municipal WWTPs have been constructed to treat urban sewage, with a combined capacity of 157 billion liters per day. In each of these facilities, sewage from tens to hundreds of thousands of individuals creates an enormous biological reactor where bacteria and resistant genes are exposed to significant concentrations of selective agents such as antimicrobial agents, disinfectants, and heavy metals. In this respect, resistant genes detected in sewage can be seen to represent the resistance burden of their urban populations. Therefore, resistance profiles in sewage reflect the structure and diversity of resistant bacteria in the gastrointestinal tracts of urban residents within the WWTP catchment. A nationwide survey of resistance elements in sewage (untreated influent) could then provide a rapid and efficient method for assessing the burden of antibiotic resistance of urban populations.

A team of scientists from the Institute of Urban Environment, Chinese Academy of Sciences, and The University of Hong Kong, conducted a nationwide investigation in China to address the profiles of all antibiotic-resistant genes (resistome) with seasons and regions. In their study, a large-scale sampling of municipal sewage from 17 major cities across China was performed. In total, 116 urban sewage samples were collected from 32 WWTPs during summer and winter, and sampling sites were specifically chosen to reflect diverse climatic conditions, economic development levels and urban geography. By combining metagenomics analyses and illumina sequencing of 16S ribosomal RNA genes, the seasonal variation and geographical distribution of the urban sewage antibiotic resistome were characterized (Fig. 1).

This study revealed that municipal sewage harbored diverse and abundant resistance genes. In total, 381 different resistance genes conferring resistance to almost all antibiotics were detected and these genes were extensively shared across China, with no geographical clustering, highlighting that municipal sewage could be a major conduit for transferring antibiotic resistance genes into the environment.

Seasonal variation in abundance of resistance genes was observed, with average concentrations of 3.27×10^{11} and 1.79×10^{12} copies/L in summer and winter,

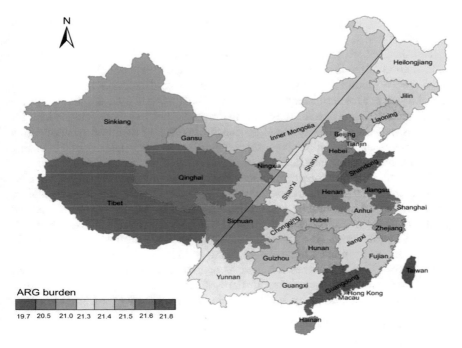

Fig. 1 Map showing the ARG burden based on urban populations of administrative districts in China. The black line on the map refers to the Chinese demographic "Hu Huanyong line" marking the difference in the distribution of China's population. Green signals a low and red a high burden of ARG

respectively. Global comparison and risk assessment are the next steps, as currently we only limited information at larger-scale information. Bacterial communities did not exhibit geographical clusters, but did show a significant distance–decay relationship, which means the patterns differ more between locations further way. The researchers also found that the core human gut microbiota was strongly associated with the shared resistome, demonstrating the potential contribution of human gut microbiota to the dissemination of resistance elements via sewage disposal. Importantly, this study observed a strong spatial dependency in the distribution of ARG abundance in various administrative areas, with two main regions separated by the demographic "Hu Huanyong line," which is based on climatic zonation and population density. This suggests that monitoring sewage systems for ARGs could provide a real-time estimate of antibiotic resistance threats in specific areas, and this in turn could be used to inform treatments and promote antibiotic stewardship.

Detection and measurement of antimicrobial resistance are essential for understanding their potential to adversely affect human health. To obtain the surveillance data of antimicrobial resistance in urban sewage at a large scale, identifying environmental reservoirs of antimicrobial resistance and pathways that pose potential health risks to human, animals, and other biota—are essential components of a systems approach to AMR.

This detailed analysis of antimicrobial resistance in Chinese urban sewage gives an indication of the magnitude of the burden imposed by resistance in China, and it should lead to increased efforts to control antibiotic resistance. Given the growing global resistance for antibiotic and documented health issues related to inappropriate use of such antibiotics, this study has major public health policy implications for countries like China. The systems' approach provides a baseline for investigating environmental dissemination of resistance elements and raises the possibility of using the abundance of resistance genes in sewage as a tool for antibiotic stewardship.

Emission of antimicrobial resistance from humans mapped for major Chinese cities revealed that urgent actions are needed to tackle the problem. New regulatory approaches to mitigate antimicrobial resistance should be developed and national and local data on antimicrobial resistance must be collected and made publicly available to enhance antimicrobial resistance surveillance systems. Antimicrobial stewardship is a critical tool for preventing and controlling antimicrobial resistance.

In August 2016, China has unveiled its national action plan for AMR. As part of a systems approach to reduce the risks of AMR, following are some actions recommended for reducing the health risks of AMR—not only for Chinese cities:

- Develop national multi-sector action plans
- Develop new antimicrobials
- Make sales of antimicrobial drugs by prescription only
- Increase awareness about AMR among the general public and policymakers
- Improve public information on the safe use of antimicrobial medicines
- Increase training and education for medical professionals and consumers on proper use of
- Build surveillance systems for antimicrobial resistance
- Enforce regulations on the sale of antimicrobial medicines without prescription
- Intensify international cooperation and exchanges to prevent and control AMR.

Acknowledgements The work was recently published in the journal Microbiome and was supported the Natural Science Foundation of China and the Chinese Academy of Sciences (CAS).

References

Science (2016) China tackles antimicrobial resistance. https://doi.org/10.1126/science.aah7247

Su JQ et al (2017) Metagenomics of urban sewage identifies an extensively shared antibiotic resistome in China. Microbiome 5(1). https://doi.org/10.1186/s40168-017-0298-y

Welcome Trust (2017) Drug-resistant infections: leading the global response, https://wellcome.ac.uk/what-we-do/our-work/drug-resistant-infections

Xiao YH, Lanjuan L (2016) China's national plan to combat antimicrobial resistanceThe Lancet Infectious Diseases, London, pp 1216–1218

Financing and Implementing Resilience with a Systems Approach in Beirut

Jieling Liu

Key Messages

1. An integrated urban environment is beneficial for health, the ultimate goal of resilient cities. It is increasingly challenged in Beirut by rapid urbanisation outpacing appropriate planning.
2. Improving health in Beirut was prioritised to focus on planning public transport, green and public spaces and walk ability. Each can be an entry point for taking a systems approach for urban resilience in Beirut.
3. Resolving this priority in Beirut can lead to improving urban health and wellbeing and achieving Roadmap 2030 for making city-regions healthy, resilient and sustainable.
4. An integrated collaborative systems modelling and implementation approach, proposed by BAU, TRUST and UHWB can improve planning to resolve the interconnected urban health problems of Beirut.
5. Harnessing complexity for resilience building in Beirut should take into account its diverse sociocultural profile and distinctive development phases from research to financing and implementation.
6. Knowledge-action transfer needs to extend from networks (KANs) to systems (KASs) and incorporate committed financing to make plans actionable for resilience in the long term. The Resilience Brokers Programme of TRUST can facilitate this process.

J. Liu (✉)
Climate Change and Sustainable Development Policies at the Institute of Social Sciences, University of Lisbon, Lisbon, Portugal
e-mail: jielingliu@campus.ul.pt

1   Urbanisation Context of Beirut

In the past five decades, Lebanon has gone through consecutive waves of rapid urban expansion driven by rural exodus, suburbanisation, war displacements and an influx of refugees. More than 87% of the country's population living in urban areas, and 64% are estimated to be residing in large agglomerations in the metropolitan areas of Beirut and Tripoli (Un-Habitat 2016). Beirut, the capital city, is facing multiple urban environmental challenges including pollution or degradation of water, soil, and air. They are pollution aggravated by climate change and demographic pressures, in the form of a high influx of refugees. The lack of green and public spaces for walking and exercise in the city and heavy reliance on private transport have been identified as some of the most obvious and urgent urban health problems to be tackled in Beirut. The city's urban planning capacity is overwhelmed by these challenges which are essentially complex.

Therefore, it has been suggested that the City of Beirut adopts a holistic and systemic vision and jointly manages urban planning incorporating all levels of institutions, practices and procedures, to issue urban policies and deliver quality services to cope with its complex and interconnected urban health challenges. Addressing issues of urban health and wellbeing in Beirut is delicate regarding the plural cultural communities in unequal development phases. Nevertheless, the variety of challenges, if addressed by a systems approach, also means high potential of encountering solutions (Lawrence and Gatzweiler 2017).

Good health and wellbeing are the most forthright indicators and essential goals of urban environmental sustainability. This message is reflected in many global and local commitments. For the City of Beirut, the Urban Resilience Masterplan resulting from a comprehensive risk assessment and planning process from 2015–2017 aims to strengthen understanding of multi-hazard risks, develop the city's preparedness and response capacity, and better support and catalyse the city-level investment plans needed to protect lives and assets.

2   April 2017 Seminar

Resonating with the same aim, Beirut-Arab University (BAU), in collaboration with the International Council for Science (ICSU) global science programme on systems thinking for urban health and wellbeing (UHWB), organised an interdisciplinary seminar entitled "Urban Health and Wellbeing: Advancing Systems, Science and Technology" during April 26–27, 2017. Participants including members and sponsors of the ICSU programme, the CEO of the Ecological Sequestration Trust (TRUST) and a group of experts in urban sciences from Beirut-Arab University, jointly identified priorities for improving health and wellbeing in the City of Beirut.

The April seminar introduced participants to the fundamentals of systems approaches, which facilitate knowledge creation and effective action through the use of computer-supported systems tools alongside participatory processes of engagement with academics, professionals and citizens. At this workshop, participants

- agreed that systems approaches are needed to effectively harness urban complexity and improve health and wellbeing in the City of Beirut
- developed a preliminary priority list of action areas to be addressed
- underscored the value of a modelling platform to illuminate urban complexity and facilitate collaborative action for implementation of sustainable development priorities
- decided to organise a workshop in October 2017 to further explore the systems approach and to establish collaborative systems modelling suitable for Beirut and
- concluded to organise an international conference on implementing systems approaches for urban health and wellbeing in 2018.

3   October 2017 Workshop

Implementation of the resilience strategy and associated action plan such as the aforementioned Urban Resilience Masterplan for the City of Beirut is expected to improve the health and wellbeing of Beirut's residents. In particular, planned investments in infrastructure, risk preparedness and recovery capacity should reduce the vulnerability of Beirut's residents to natural and man-made hazards, including the risks associated with climatic, epidemiological and demographic changes, in part by ensuring access to fresh water, sanitation and energy and enhancing mobility. These investments in resilience will also contribute to the inclusive implementation of the SDGs and NUA, most critically by improving human health and wellbeing, reducing poverty and providing affordable housing for all. The success of this broad effort to secure resilience will depend on the adoption of an integrated systems approach, a reality increasingly recognised by the scientific and policy communities.

Hence, the second workshop in October 2017 aimed to present Roadmap 2030 (TRUST), which sets out an action plan to deliver the agreed Global Goals and implement the New Urban Agenda (NUA), and the Urban Resilience Master Plan for the City-Region of Beirut. The goal was also to focus on the application of systems thinking to urban health and wellbeing challenges, including model development and application throughout the process of drafting the implementation plan for Beirut. Finally, the workshop aimed at building capacity and contributing to an international conference at BAU planned for September 2018.

Partners and participants involved in the October 2017 workshop

- Beirut-Arab University (BAU), represented by a group of experts in urban sciences, with the prospect of establishing a centre of urban health.
- International Council for Science (ICSU) global science programme on systems science for urban health and wellbeing (UHWB), represented by its executive director and members of the programme's scientific committee.
- Ecological Sequestration Trust (TRUST), represented by its CEO and Chief of Platform Delivery.

4 Finding 1: Building Resilience by Applying a Systems Approach

The urban and environmental contexts in which problems of public health and well-being emerge are interconnected and complex. Consequently, implications on health impacts and quality of life can be costly. It is important to understand the multiple layers of economic, cultural, political and ecological elements as well as stakeholders that compose a city, prior to offering humanitarian or developmental solutions. This urban complexity can be properly captured and modelled using a systems approach: co-producing knowledge for urban health and wellbeing in collaboration with science.

As Beirut is facing multiple complex urban environmental challenges, using a systems approach can help Beirut to better harness urban complexity, facilitate data-informed transdisciplinary intelligence to synergize urban planning and policymaking at different levels among diverse stakeholders and hence generate solutions that could build resilience for Beirut.

5 Proposed Architecture for Harnessing Complexity and Building Resilience

The participants of the second Beirut workshop on "Implementing Resilience in the City-Region of Beirut" elaborated and demonstrated an approach to enhancing health and wellbeing. The architecture proposed to harness complexity for resilience building includes four equally significant components:

- Evidence-based, transdisciplinary knowledge as a result of taking a systems approach across sectors and agents in urban systems.
- A systems approach, which can guarantee the transdisciplinarity in the process of data gathering and that of involving stakeholders.
- An integrated tool of analysis comprising collaborative modelling and evaluation to produce scientific outcomes and indicate gaps for actions in policymaking and financing.
- Adequate financing models ensure the shift from knowledge to action, provide incentives and sustain transformations (Fig. 1).

Fig. 1 Proposed architecture to harness complexity for resilience building

6 Finding 2: Resilience Brokers Programme—a Collaborative Model for Healthy and Resilient City-Regions

The Ecological Sequestration Trust (TRUST) has brought together a multidisciplinary group of global leading experts and organisations, the Resilience Brokers, to develop and implement the Resilience Brokers Programme, an initiative designed to facilitate the implementation of Roadmap 2030, an action plan for achieving the SDGs, the NUA and suitable for implementing the Urban Resilience Masterplan for the City of Beirut.

The programme champions a collaborative human-ecological-economic and resource systems approach (CHEER approach) to harness data and scientific evidence for investment and planning decision making by integrated modelling of social and natural systems and their interlink ages. The programme aims to foster new forms of collaboration between the public and private sectors by acting as a neutral catalyst and to promote new and effective means of involving communities and building their capacity for innovation.

This ambitious initiative aims to support the financing of sustainable development paths in 200 global city-regions by 2022, to trigger a rapid scale up to 70% of all city-regions in the world by 2030 and to improve the lives of more than five billion people. The programme's high potential for global impact underpinned by the advanced technology of resilience.io has attracted leading organisations from all sectors of society. Resilience Brokers Programme delivery partners include the Group on Earth Observations (GEO), the International Centre for Earth Simulation (ICES), the Urban Climate Change Research Network (UCCRN), Imperial College London (ICL), the Institute for Integrated Economic Research (IIER), the United Nations Sustainable

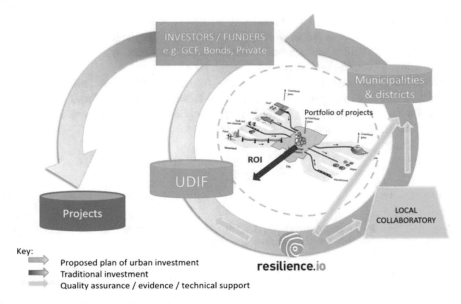

Fig. 2 The resilience brokers programme

Development Solutions Network (UNSDSN) and the International Council for Local Environment Initiatives (ICLEI). The combined knowledge, far-reaching networks, influencing power and established operational excellence of the Programme partners allows for rapid mobilisation and seamless deployment of the Programme in any region in the world—making the vision of its ambitious scale-up schedule a practical and achievable goal.

The Resilience Brokers Programme as a unique opportunity for government and business to invest in leading-edge solutions is proposed to the City of Beirut to build resilience and to promote urban health and wellbeing, with the priority set on improving the city's green and public spaces and walkability (Fig. 2).

7 Policy Recommendations

1. A systems approach is recommended to harness the complexity of urban health challenges, to analyse existing policy and financing gaps and to identify priorities for actions that can plan-out risk and build-in resilience.
2. A systems approach is recommended to co-create collective intelligence from multiple stakeholders of public and private sectors, civil society organisations and communities, in order to foster social learning and make policies more inclusive for community resilience capacity building.

3. Policies should facilitate knowledge-action networks to be extended into knowledge-action systems, which is important to enable the functioning of urban systems for health and secure stable financial resources and thereby making resilience-building plans actionable in the long term.
4. An integrated science–policy–financing framework of collaborative systems modelling and implementation, such as the Resilience Brokers Programme, has come of age in the issue of addressing complex urban health and wellbeing problems, for instance, in the City of Beirut.

References

About the Urban Resilience Masterplan for the City of Beirut: "Urban Resilience Masterplan for the City of Beirut" Available at: https://nl4worldbank.org/2015/05/29/urban-resilience-masterplan-for-the-city-of-beirut/
"EMI Beirut Mission" Available at: http://emi-megacities.org/news/emi-beirut-mission/
Gatzweiler FW, Ayad HM, Boufford JI, Capon A, Diez Roux AV, Donelly Ch, Hanaki K, Jayasinghe S, Nath I, Parnell S, Rietveld L, Ritchie P, Salem G, Speizer I, Zhang Y, Zhu Y-G (2016) Advancing urban health and wellbeing in the changing urban environment. In: Implementing a Systems Approach. Zhejiang University Press, Hangzhou and Springer, Singapore
Government of Lebanon, Council for Development and Reconstruction (CDR), Grand Serail—Beirut, Lebanon. January 2016. Habitat III National Report
Lawrence RJ, Gatzweiler FW (2017) Wanted: a transdisciplinary knowledge domain for urban health. J Urban Health https://doi.org/10.1007/s11524-017-0182-x
The Ecological Sequestration Trust. Peter Head CBE FREng 2016. Roadmap 2030: Financing and Implementing the Global Goals in Human Settlements and City-Regions. Version 1.0. Available at: https://ecosequestrust.org/roadmap2030/
What Is Resilience.io? Available at https://resilience.io/resilience-io-supported-by-the-ecological-sequestration-trust/

Addressing the Environment and Health Nexus is a Strategic Approach to Advance the Sustainable Development Goals in ASEAN

M. J. Pongsiri and V. Arthakaivalvatee

Key Messages

1. The Southeast Asia region is at risk of losing hard won health and development gains due to the global and local environmental changes we are now experiencing.
2. The Association of Southeast Asian Nations (ASEAN) is the opportune regional platform to actively engage at the environment and health nexus as a strategic approach to deliver on the Sustainable Development Goals (SDGs).
3. ASEAN is committed to delivering on the SDGs through coordination, information sharing and reporting, with Thailand as the designated ASEAN lead country.
4. The ASEAN Declaration on Culture of Prevention (CoP) for a Peaceful, Inclusive, Resilient, Healthy and Harmonious Society is essential to how we address the environment and health nexus in the region because our increased attention to the environmental determinants of health presents new opportunities to improve the environment, to improve health, and to prevent adverse health risks in ASEAN communities.
5. It is time to work with ASEAN to help address the region's linked environment and health priorities by demonstrating the use of knowledge-based tools for integrated assessment, monitoring, modelling and valuation—all priority policy needs for demonstrating the value of addressing environment and health together (the "environment and health nexus") rather than as separate sectoral issues.

M. J. Pongsiri (✉)
Cornell University, New York, USA
e-mail: mjp329@cornell.edu

V. Arthakaivalvatee
Thailand Institute of Justice (TIJ), Bangkok, Thailand
e-mail: Vongthep@asean.org

© Zhejiang University Press and Springer Nature Singapore Pte Ltd. 2020   57
F. W. Gatzweiler (ed.), *Urban Health and Wellbeing Programme*,
Urban Health and Wellbeing, https://doi.org/10.1007/978-981-15-1380-0_10

Fig. 1 WHO estimated that in 2012, 23% of all deaths worldwide (12.6 million people in total) are attributed to environmental factors. 30% of all environmentally related deaths occur in Southeast Asia

Human health and well-being depend on the state of our environment (Prüss-Üstün et al. 2016). Unhealthy environments are linked to at least 23% of global deaths (Prüss-Üstün et al. 2016). Understanding the environmental determinants of health can provide opportunities and practical strategies to reduce and prevent risks to health (Prüss-Üstün et al. 2016) and to improve the state of the natural systems on which we depend (Whitmee et al. 2015).

The Southeast Asia region is at risk of losing hard won health and development gains due to the global and local environmental changes we are now experiencing (UNESCAP 2015).The region accounts for 30% of global deaths due to environmental factors (Fig. 1) (Prüss-Üstün et al. 2016). These are preventable deaths. The region has environment and health hotspots due to its vulnerability to climate change, infectious disease risk, increasing risk of non-communicable diseases (NCDs) and need for both strengthened health systems and environmental management practices.

Strong partnerships and understanding among key stakeholders will be necessary to identify and implement sustainable solutions to challenges at the environment and health nexus such as climate change and heat stress, the human toll of extreme weather events, water scarcity, food security, vector-borne diseases and non-communicable diseases related to unhealthy lifestyles in our built environments.

It is now time to actively engage the region at the environment and health nexus as a strategic approach to deliver on the Sustainable Development Goals (SDGs). Such an approach could measurably benefit both human health and the environment (Dora et al. 2015). There is already an existing regional platform to operationalize such co-benefits-based activities—The Association of Southeast Asian Nations (ASEAN). ASEAN is a regional intergovernmental organization that includes the following member states: Indonesia, Thailand, Malaysia, Singapore, the Philippines, Brunei, Vietnam, Laos PDR, Myanmar and Cambodia (Fig. 2). ASEAN aims to accelerate economic growth, social and cultural development in the region through cooperation on shared interests and challenges.

Guided by the ASEAN Vision 2025, ASEAN's ambitious agenda for achieving an integrated community is through the implementation of its political-security, economic and socio-cultural blueprints (ASEAN Community Vision 2025 2018).

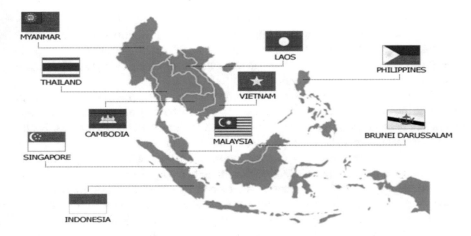

Fig. 2 ASEAN Member Countries[8]

As a regional grouping, in complementation with its member countries' sovereign responsibilities, ASEAN is also committed to delivering on the 2030 Agenda for Sustainable Development through coordination, information sharing and reporting, with Thailand as the designated ASEAN lead country.

1　Policy Conditions Are Favourable to Engage ASEAN Now

It is time to work with ASEAN to help address the region's linked environment and health priorities by demonstrating the use of knowledge-based tools for integrated assessment, monitoring, modelling and valuation—all priority policy needs for demonstrating the value of addressing environment and health together (the 'environment and health nexus') rather than as separate sectoral issues. However, there is a need to strengthen capacity in ASEAN to work at this nexus, using an *integrated* approach which considers development, environmental, and social concerns; and, based on the best available evidence to inform policies to promote, plan for, and implement sustainable development.

In a ground-breaking move, ASEAN leaders adopted the ASEAN Declaration on Culture of Prevention for a Peaceful, Inclusive, Resilient, Healthy and Harmonious Society in 2017. Similar to the UN Secretary General's prevention agenda, the Culture of Prevention (CoP) was developed to address the root causes of socio-economic issues in the ASEAN region, including varying forms of violence, environmental degradation and quality of life. Prominent among its six substantive thrusts are: (i) thrust 4: promoting the culture of resilience and care for the environment and (ii) thrust 5: promoting the culture of healthy lifestyle (ASEAN 2018).

The CoP is by and large a paradigm shift in its own right. It not only provides coherent policy direction to ASEAN Member States to mainstream a preventive approach across all of ASEAN's core pillars of work—political-security, economic and socio-cultural, but also a platform to foster a mindset change among their citizens from a reactive to preventive one.

This policy development is essential to how we address the environment and health nexus in the region because our increased attention to the environmental determinants of health presents new opportunities to improve the environment, to improve health and to prevent adverse health risks in ASEAN communities.

The growing evidence of significant harms to health and well-being from environmental changes in the ASEAN region, together with Thailand's chairmanship of ASEAN and the theme 'Advancing Partnership for Sustainability' in 2019, provide fortuitous timing and opportunity for key stakeholders—the scientific community, civil society, and policymakers-to actively engage with ASEAN at the environment and health nexus to improve health and well-being, the prerequisites to human development and long-term economic progress.

The United Nations (UN) share an interest in addressing environmental sustainability and human health together—not just increasing understanding of critical environment and health relationships but also applying that understanding to inform policy through integrated assessments, identification and analyses of policy interventions to improve both health and the environment. Integration underlies the Sustainable Development Goals (SDGs)—so, adopting an approach to exploit synergies among the SDGs and minimize trade-offs to planning and implementing activities is practical but also a challenge to existing governance structures. **A system-based understanding of environment and health nexus challenges, including their causes (singly or in combination, direct or indirect) and feedback loops (which affect them positively or negatively), is critical to identifying preventive policy strategies and cost-effective solutions for ASEAN to deliver on the 2030 Agenda**.

With the UN now a formal partner with ASEAN, a renewed focus, and cross-UN partnering on the environment and health nexus (UNEP 2015), the SDGs are the common framework to apply an integrated, policy coherent approach.

Importantly, there is agreement that we can stay on a positive sustainable development pathway if the human activity drivers, and health and well-being consequences, of global and local environmental changes are understood and this understanding is reflected in policy and planning (Whitmee et al. 2015).

2   Starting with a Strong Science Base

A more rigorous science base is needed to plan for and implement the SDGs particularly those related to health (SDG 3), safe water (SDG6), safe cities (SDG 11) and climate action (SDG 13). A recent report concluded that the Asia-Pacific region is making slow, insufficient progress on the SDGs, with no/little progress on terrestrial ecosystems (SDG 15), and climate action, largely attributed to air pollution. And, the report findings reinforced the need for a more integrated and inclusive approach to produce the data needed to underlie planning for and measuring SDGs (ESCAP 2017). ASEAN recognizes the challenges to addressing the environment and health nexus in practice, and importantly, the critical need to adopt approaches which require information sharing and policy coordination among environment, health, and finance sectors (UNEP 2015).

Applying a system-based analysis to build a shared understanding of environment and health nexus challenges can also help to assess the available evidence for the development of integrated policy tools to address scientific research priorities such as how understanding of dynamic interrelationships between social, economic and environmental factors (the "system") helps identify or address trade-offs and unintended consequences of policy choices (Pongsiri 2018). Such policies are developed at multiple levels—national, subnational and local. At all levels, decision-makers must set priorities among interconnected, and sometimes competing, issues. A system-based understanding of sectoral challenges, particularly the dynamic relationships between drivers and consequences across sectors, over spatial and temporal scales, should be the basis for identifying priorities within a specific context. When interconnections and feedbacks are not recognized at the time of policy formulation, the opportunity for mitigation of adverse health, environmental, and distributional, effects as well as for capitalizing on co-benefits to advance multiple objectives, is lost (Johnston et al. 2012).

3   Case Study: Human Health Impacts of Landscape Fires

Landscape fires cause approximately 30,000 premature deaths every year (Koplitz et al. 2016). They are also major drivers of biodiversity loss, particularly in Southeast Asia where setting fires to clear land for agriculture is common practice. It is estimated that approximately 100,000 deaths across Singapore, Malaysia and Indonesia were attributed to the fires in the year 2015. These are deaths which otherwise would not have happened if not for the fires. 2015 was an El Nino year which created drought conditions, drying out the organic-rich peatlands where most of the fires were set. A multidisciplinary team of scientists studied the relationship between land use, fire emissions associated with land cover change, their wind-driven transport towards population centres, and what we know epidemiologically about the human health effects of exposure to particulate matter (PM2.5) in fire emissions. The integration of these environmental and health data across disciplines resulted in the development of an innovative modelling tool which quantifies the human health impacts (mortality) of fire events. Quantified human health costs could be used to justify policy decisions aimed at preventing the greatest human health risks. Moreover, the modelling tool could be adapted to inform country and local-level policies to help address seasonal fires through the protection of peatlands.

The use of the integrated modelling tool to position health as a primary concern strengthens the evidence base to help prioritize peatlands protection planning so that where fires avoided, the greatest human health risks could be prevented downwind. In addition, the same health and economic costs and benefits of peatlands protection could support the evidence base for climate change mitigation efforts. A system-based understanding of the drivers and consequences of land use clearing by fire for agricultural production helped to identify leverage points for intervention to improve health as well as opportunities to take a co-benefits-based approach to protect peatlands and reduce risks to human health. Active engagement with decision-makers reinforced interest in the tool's early warning potential.

4 Policy Recommendations for Addressing the Environment and Health Nexus in ASEAN

- Before policy formulation, recognize the interconnections and feedbacks of the environment-health nexus challenge so as not to lose the opportunities for mitigation of adverse health, environmental, and distributional, effects as well as for capitalizing on co-benefits to advance multiple objectives.
- Highlight the environment-health nexus in the implementation of the ASEAN Vision 2025 and the 2030 Agenda for Sustainable Development, especially their complementarities in doing so.
- Engage key stakeholders to be key partners in promoting the environment-health nexus, and particularly in incentivizing the private sector.
- Promote broader public understanding of the environment-health nexus to increase public acceptance and support.
- Identify the environment and health benefits of taking integrated, co-benefits-based policy action on the environment-health nexus challenges. Identify the environment and health costs of no policy action.
- Generalize how lessons learned to address the environment-health nexus in a specific context could apply to other places experiencing (or expecting) similar challenges.

References

ASEAN Community Vision 2025. Available at: https://www.asean.org/wp-content/uploads/images/2015/November/aec-page/ASEANCommunity-Vision-2025.pdf. Accessed 16 Dec 2018

ASEAN Countries https://www.worldatlas.com/articles/asean-countries.html. Accessed 16 Dec 2018

ASEAN declaration on culture of prevention for a peaceful, inclusive, resilient, healthy and harmonious society (2.iv–2.v.)

Complementarities between the ASEAN Community Vision 2025 and the United Nations 2030 Agenda for Sustainable Development: A Framework for Action. 2015. https://www.unescap.org/publications/complementarities-between-asean-vision-2025-and-2030-agenda

Dora C et al (2015) Indicators linking health and sustainability in the post-development 2015 agenda. Lancet 385:380–391

ESCAP (2017) Asia and the pacific sdg progress report. United Nations, Bangkok

Johnston FJ et al (2012). Estimated global mortality attributable to smoke from landscape fires. Environ Health Perspect 120:695–701

Koplitz S et al (2016) Public health impacts of the severe haze in Equatorial Asia in September–October 2015: demonstration of a new framework for informing fire management strategies to reduce downwind smoke exposure. Environ Res Lett 11(9)

Osofsky SA, Pongsiri MJ (2018) Operationalising planetary health as a game-changing paradigm: health impact assessments are key Lancet Planet Health 2(2): e54–55

Whitmee, Safeguarding Human Health. Stokstad (2015) Sustainable goals from U.N. underfire. Science 347(6223):702–3

UNEP/APEnvForum/3, 2015

UNEP/EA.3/L.8/Rev.1; Global action plan for healthy lives and well-being for all. http://www.who.int/sdg/global-actionplan

Whitmee S et al (2015) Safeguarding human health in the Anthropocene Epoch: report of the Rockefeller foundation-Lancet commission on planetary health. The Lancet 386(10007):1973–2028

Prüss-Üstün A, Wolf J, Corvalán C, Bos R, Neira M (2016) Preventing disease through healthy environments: a global assessment of the burden of disease from environmental risks. World Health Organization. https://apps.who.int/iris/handle/10665/204585

WHO Environmental Impacts on Health. http://www.who.int/quantifying_ehimpacts/publications/PHE-.prevention-diseases-infographic-EN.pdf. Accessed 16 Dec 2018

Health and Well-Being in the Changing Urban Environment

Xinjue Ke and Franz W. Gatzweiler

Key Messages

1. Integrating health indicators into all policies and implementing integrated systems governance could effectively address multiple change challenges by creating health co-benefits.
2. The guiding principles of the urban planning system need to be inclusive of health.
3. Cities' capacity to deal with emergent public health issues can be enhanced by integrating health into all policies.
4. Public participation and community capacity building for urban health need promotion.
5. Enhance research and education on healthy cities.
6. Set local goals and assess progress indicators towards health goals regularly.

1 Policy Context

The world is undergoing a severe wave of urban population growth and sprawl, driven by economic growth, population increase and migration. In 2018, approximately 55% of the population settled in urban areas (United Nations 2018), and by 2050, 70% of the world's population is projected to be urban (WHO 2018). Today, Latin America

X. Ke (✉) · F. W. Gatzweiler
Programme on Urban Health and Wellbeing, the Global Interdisciplinary Science Programme on Urban Health and Wellbeing, Xiamen, China
e-mail: xjke@iue.ac.cn

F. W. Gatzweiler
e-mail: gatzweiler02@gmail.com

© Zhejiang University Press and Springer Nature Singapore Pte Ltd. 2020
F. W. Gatzweiler (ed.), *Urban Health and Wellbeing Programme*,
Urban Health and Wellbeing, https://doi.org/10.1007/978-981-15-1380-0_11

and the Caribbean region have 81% of its population dwelling in urban areas. Asia has approximately 50% and Africa, even though it is still mostly rural, is embracing fast urban sprawl over the next decade (United Nations 2018).

Urbanisation as a global concern is closely linked with multiple disciplines, including urban planning, economics, sociology and geography, which directly or indirectly affect human health. A set of major public health matters such as emerging non-communicable and communicable diseases, health inequalities, climate change and job insecurity have arisen as severe consequences of urbanisation. In 2015, the Chinese government launched the 'Belt and Road Initiative' which provides fundamental to achieve global health reform. In 2017, the World Health Organisation and China have entered a strategic partnership for establishing the 'Healthy Silk Road'. Health is embedded in the investment in and development of infrastructures and transportation in participating countries as main strategies to improve public health. 'Healthy China 2030' as a national strategy defined the health sector as a priority for global sustainable development, promoting good health for all. It has recognised that economic development goes hand-in-hand with a healthy population and environment.

Significant actions have been taken to improve the health of cities globally. In 2016, the New Urban Agenda launched by UN Habitat, put health as a central concern of urban planning. UN Habitat started a working relation on health and planning with WHO and the Shanghai Consensus on Healthy cities was attended by more than 100 mayors from around the world. The mayors were committed to good governance for health, with core principles of considering integrated health into all policies and addressing all health determinants from social, economic and environmental aspects (Healthy Cities Mayors Forum 2016).

Nevertheless, the health sector alone is not able to perform well in addressing these concerns. The development of a collaborative and integrated approach is needed to understand and address complex urban health issues. Rather than an outcome, a healthy city is more a process that continually attempts to improve the health of its residents (WHO 2018a, b).

Over the past few years, system approaches to urban health and well-being in Asia, the Pacific, Latin America, the Caribbean and Africa have been improving the regions' understanding of urban complexity, particularly its effect on health. In Malaysian cities, a Salud Urbana en America Latina (SALURBAL) project draws on the collaborative conceptual modelling (CCM) approach, promoting more effective decision-making for urban health and sustainability. In El Salvador, the 'Urban Health Model', a participatory decision-making process for 'Health in All Policies' informed by participatory systems modelling, has been implemented. It generates information and knowledge from different urban sectors and from different types of data to formulate policies and actions in response to emerging threats to urban health and well-being.

2   International Symposium

In October 2018, the International Council for Science's (ICS) global science programme on Urban Health and Well-being (UHWB), in collaboration with the Institute of Urban Environment and the Chinese academy of Science (CAS), organised an International Symposium on Health and Well-being in the Changing Urban Environment. It assembled more than 30 experts in relevant fields from eight countries all around the world. The symposium aimed at providing an opportunity to present evidence and case studies on the possible generated benefits by better understanding the interconnectedness of specific urban health and well-being issues and addressing them by taking systems approaches and by promoting health as a core of urban policy-making.

In line with the WHO's goal of promoting the collaboration between different urban sectors including transportation, housing, education, economy, etc., as well as fostering community participation and maximising the effectiveness of local governance, the symposium identified several key areas required for improving urban health:

1. **Urban Governance and planning**: Urban planning and governance are important determinants of urban health. Health, as one of the most important sectors within the urban system, should be considered at the beginning of the planning and policy-making process rather as an expected outcome. Cities are rapidly growing and developing into more dynamically complex systems, which produce large amounts of unintended urban health issues that have an impact on residents' quality of life. Solely promoting programmes on improving urban health could be ineffective as they are unable to maximise multi-urban system functions and generate co-benefits. In contrast, territorial and spatial planning emphasises integrated urban governance, and cross-sectoral management of resources. This type of planning addresses health issues by considering, as well as engaging with, stakeholders at all levels to act across boundaries for both transitional and regional cooperation.

   Amid the era of urban transformation due to urban growth, governance requires increasing collaboration across disciplines. Engagement with society at all levels holds great potential to address dynamic interactions between people, ecology, and technology in cities. Urban residents' lifestyles, which are less physically active, high fat diets and with high levels of psychological pressure, make multi-level efforts to engage and call for public participation to improve public health status and well-being, indispensable and inevitable.

2. **Technologies and networks**: Cities are complex systems, having multiple interacting networks which contribute to the emergence of good health and well-being. Strengthening collaborative networks, that comprise of people, artefacts or machines, is necessary for addressing urban health issues. This is due to the fact that health is an "emergent property" resulting from different interactions among components of a complex and adaptive system.

   As part of the networks, innovative and interactive technologies include big data technologies, Geographic Information Systems (GIS) and geo-detectors. These are important tools for harnessing the complexities of urban health issues and have great potential to promote healthy cities such as eco-city and smart city development. In the future, universities should play a major role in fostering, educating and strengthening a variety of forms of such networks and interactions.

3. **Communication and research design**: Population growth driving urbanisation leads to a series of significant changes in urban settings of education, commerce and economy. An integrated and participatory research design to adapt and address urban complexity is required as cities are facing unprecedented challenges. Through visualising difficult concepts, design itself is able to build the bridge between disciplines. It also builds bridges between rapid progresses in technology, science and engineering. The approaches towards shaping the relationship between health, well-being and urban environment, therefore, become vital.

4. **Measuring urban pollution and environmental risk**: Vulnerable groups in cities, particularly women, children and the elderly, are more vulnerable to environmental pollution. Today, more than 93 per cent of children all around the world breathe severely polluted air (WHO 2018a, b). Urban built environments, including industrial land, green and open space, and road systems are closely linked to respiratory health. That should be considered in the strategies for healthy urban planning. Furthermore, understanding the effectiveness of a range of interventions in different sectors such as housing, planning and transport could contribute to air and water quality improvement.

   Risk assessments of public health impacts of the climate change are important for policy selection, which contribute to a risk-informed decision-making processes. Many cities are located in areas prone to natural disaster and climate hazard. The health of city dwellers is expected to be affected by direct physical injuries, water-borne diseases and respiratory illnesses due to changing weather conditions. An analysis of such vulnerabilities at both spatial and temporal scale is able to help identifying policy options considering feasibility, applicability and robustness.

5. **Housing and health**: In both developing and developed countries, housing and the built environment affect residents' physical and mental health and social well-being. The World Health Organization (WHO) recently published guidelines which interpreted 'Healthy Housing' as one that provides a safe and healthy environment for its residents (WHO 1988). Given the situation of growing inequalities in access to quality housing and affordable energy, taking system approaches which underpin the United Nation Sustainable Goals could improve renewable energy to produce affordable housing at WHO minimum temperatures.

3   Conclusion

Cities offer residents numerous opportunities to access jobs, goods and community services, as well as create opportunities for health. However, the world has a huge population with a dramatic increase in urban dwellers, which exacerbates the sources of social and environmental strains which adversely affect human health and well-being in cities. The ultimate challenge for today's urban expansion is to seek a way of maximising the functions of a multi-functional urban system and generating co-benefits to improve urban health. The international symposium reflected on the urgent demands for new and intelligent urban planning strategies. Integrated systems governance across urban sectors is a promising strategy to improve health and well-being in cities. In order to achieve health goals for all, an attempt is required to integrate health in all policies and better understand the complex interactions between urban health, well-being and the changing urban environment.

References

Healthy Cities Mayors Forum (2016) Shanghai consensus on healthy cities 2016. Retrieved from https://www.who.int/healthpromotion/conferences/9gchp/9gchp-mayors-consensus-healthy-cities.pdf?ua = 1

United Nations (2018) 68% of the world population projected to live in urban areas by 2050, Says UN. Retrieved from https://www.un.org/development/desa/en/news/population/2018-revision-of-world-urbanization-prospects.html

WHO (1988) Guidelines for healthy housing. Retrieved from http://apps.who.int/iris/bitstream/handle/10665/191555/EURO_EHS_31_eng.pdf?sequence=1

WHO (2018a) Retrieved from healthy cities: http://www.wpro.who.int/china/mediacentre/factsheets/healthy_cities/en/

WHO (2018b). More than 90% of the world's children breathe toxic air every day. Retrieved from https://www.who.int/news-room/detail/29-10-2018-more-than-90-of-the-world's-children-breathe-toxic-air-every-d

Printed in the United States
By Bookmasters